THE U.S. NAVAL INSTITUTE ON
NAVAL STRATEGY

U.S. NAVAL INSTITUTE
WHEEL BOOKS

In the U.S. Navy, "Wheel Books" were once found in the uniform pockets of every junior and many senior petty officers. Each small notebook was unique to the Sailor carrying it, but all had in common a collection of data and wisdom that the individual deemed useful in the effective execution of his or her duties. Often used as a substitute for experience among neophytes and as a portable library of reference information for more experienced personnel, those weathered pages contained everything from the time of the next tide, to leadership hints from a respected chief petty officer, to the color coding of the phone-and-distance line used in underway replenishments.

In that same tradition, U.S. Naval Institute Wheel Books provide supplemental information, pragmatic advice, and cogent analysis on topics important to all naval professionals. Drawn from the U.S. Naval Institute's vast archives, the series combines articles from the Institute's flagship publication *Proceedings*, as well as selections from the oral history collection and from Naval Institute Press books, to create unique guides on a wide array of fundamental professional subjects.

THE U.S. NAVAL INSTITUTE ON
NAVAL STRATEGY

EDITED BY THOMAS J. CUTLER

NAVAL INSTITUTE PRESS
Annapolis, Maryland

Naval Institute Press
291 Wood Road
Annapolis, MD 21402

© 2015 by the U.S. Naval Institute
All rights reserved. No part of this book may be reproduced or utilized in any form or by any means, electronic or mechanical, including photocopying and recording, or by any information storage and retrieval system, without permission in writing from the publisher.

Library of Congress Cataloging-in-Publication Data
The U.S. Naval Institute on naval strategy / edited by Thomas J. Cutler.
 pages cm. — (U.S. Naval Institute wheel books)
 Includes bibliographical references and index.
 ISBN 978-0-87021-113-3 (alk. paper) — ISBN 978-1-61251-888-6 (ebook) 1. Naval strategy. 2. United States. Navy—Officers' handbooks I. Cutler, Thomas J., date, editor of compilation.
 V163.U54 2015
 359.40973—dc23

2015012376

♾ Print editions meet the requirements of ANSI/NISO z39.48–1992 (Permanence of Paper). Printed in the United States of America.

23 22 21 20 19 18 17 16 15   9 8 7 6 5 4 3 2 1
First printing

CONTENTS

Editor's Note ix

Introduction 1

1 "Presence" 7
   Lieutenant Commander Thomas J. Cutler, USN (Ret.)

2 "Missions of the U.S. Navy" 10
   Vice Admiral Stansfield Turner, USN

3 "National Policy and the Transoceanic Navy" 24
   Samuel P. Huntington

4 "Make the Word Become the Vision" 45
   Captain Peter M. Swartz, USN, and
   Captain John L. Byron, USN

5 "Crafting a 'Good' Strategy" 53
   Lieutenant Colonel Frank Hoffman, USMCR (Ret.)

6 "The Global Maritime Coalition" 65
   Lieutenant Commander James Stavridis, USN

7 "Why 1914 Still Matters" 76
   Norman Friedman

8	"The Strategy of the World War and the Lessons of the Effort of the United States"..................87 *Captain Thomas G. Frothingham, USN*	
9	"Getting Sea Control Right"............................102 *Milan Vego*	
10	"Cede No Water: Strategy, Littorals, and Flotillas".....111 *Captain Robert C. Rubel, USN (Ret.)*	
11	"On Maritime Strategy"................................120 *Captain J. C. Wylie Jr., USN*	
12	"Beyond the Sea and Jointness"........................141 *Captain Sam J. Tangredi, USN*	
13	"Win Without Fighting"................................151 *Lieutenant David A. Adams, USN*	
14	"Notes on Strategy"...................................161 (Selection from the Appendix to the Classics of Sea Power edition of *Some Principles of Maritime Strategy*) *Sir Julian Corbett*	
15	"Requiem for Strategic Planning?".....................183 (Selection from Chapter 12 of *The Accidental Admiral: A Sailor Takes Command at NATO*) *Admiral James Stavridis, USN (Ret.)*	
	Index 191	

EDITOR'S NOTE

Because this book is an anthology, containing documents from different time periods, the selections included here are subject to varying styles and conventions. Other variables are introduced by the evolving nature of the Naval Institute's publication practices. For those reasons, certain editorial decisions were required in order to avoid introducing confusion or inconsistencies and to expedite the process of assembling these sometimes disparate pieces.

Gender

Most jarring of the differences that readers will encounter are likely those associated with gender. Many of the included selections were written when the armed forces were primarily a male domain and so adhere to purely masculine references. I have chosen to leave the original language intact in these documents for the sake of authenticity and to avoid the complications that can arise when trying to make anachronistic adjustments. So readers are asked to "translate" (converting the ubiquitous "he" to "he or she" and "his" to "her or his" as required) and, while doing so, to celebrate the progress that we have made in these matters in more recent times.

Author "Biographies"

Another problem arises when considering biographical information of the various authors whose works make up this special collection. Some of the selections included in this anthology were originally accompanied by biographical information about their authors. Others were not. Those "biographies" that do exist vary a great deal in terms of length and depth, some amounting to a single sentence pertaining to the author's current duty station, others consisting of several paragraphs that cover the author's career. Varying degrees of research—some quite time consuming and some yielding no results—are required to find biographical information from other sources. Because of these uneven variables, and because as a general rule we are more interested in what these authors have to say more than who they are or were, I have chosen to even the playing field by foregoing accompanying "biographies." Relevant biographical information has been included in some of the accompanying commentaries, however.

Ranks

I have retained the ranks of the authors *at the time of their publication*. As noted above, some of the authors wrote early in their careers, and the sagacity of their earlier contributions says much about the individuals, about the significance of the Naval Institute's forum, and about the importance of writing to the naval services—something that is sometimes underappreciated.

Deletions

Most of the articles included here are intact, appearing as they originally did in their entirety, but in a few cases some portions have been removed because they make suggestions or challenge policies/programs that no longer exist. Where these deletions have occurred, the following has been inserted: [...]

In the interest of space, and because *Wheel Books* are intended as professional guides, not academic treatises, citations have been removed from some of the entries.

Other Anomalies

Readers may detect some inconsistencies in editorial style, reflecting staff changes at the Naval Institute, evolving practices in publishing itself, and various other factors not always identifiable. Some of the selections will include citational support, others will not. Authors sometimes coined their own words and occasionally violated traditional style conventions. *Bottom line:* with the exception of the removal of some extraneous materials (such as section numbers from book excerpts) and the conversion to a consistent font and overall design, these articles and excerpts appear as they originally did when first published.

INTRODUCTION

While teaching the Naval War College's "Strategy and War" course for more than two decades, I have observed that the students, particularly those in military service, often come to the course looking for "the gouge," a quick checklist of things to do to win wars. This is understandable, given the military training culture, where checklists are effectively used to accomplish seemingly miraculous things, such as bringing a giant warship safely into harbor or harnessing the power of the atom. It is also a logical and understandable desire, since such a thing would make developing strategy and winning wars less daunting and more easily achievable.

In most of their endeavors, military people tend to be more scientists than artists or philosophers. We use spherical trigonometry to navigate, physics (ballistics) to deliver firepower to the enemy, thermodynamics to accomplish propulsion, and electronics to communicate, reconnoiter, and so on. All of these are exact sciences. But the granddaddy of all these endeavors is war itself, with its major components of strategy, operations, and tactics, and while there is a substantial amount of science in tactics and operations, strategy is less cooperative.

Those who have studied and/or experienced war and learned its hard lessons are very much aware that strategy is far too complex to yield to the simplicity and predictability of a checklist. So, to serve as an antidote to this checklist mentality, the War College employs a combination of theoretical readings and relevant case studies, the former serving to take the students out of the laboratory and into the lyceum where philosophy replaces scientific method, and the latter providing real-world opportunities to test the theories offered by philosophers of war.

Just as the Naval War College has grappled with this complex, elusive, and vitally important subject since its beginning, so has the Naval Institute. No one has measured the percentage of articles dealing with strategy that have appeared in *Proceedings*, but it has to be substantial. Since the first issues appeared nearly a century and a half ago, naval officers and others have tried to clarify this naturally opaque subject. Books from Naval Institute Press and the wisdom captured in the Institute's oral history program have also dealt with this challenge.

In the pages of this anthology naval professionals will find a variety of approaches to naval strategy that will edify in some cases and evoke further questions in others. These chosen pieces reveal an impressive array of erudite thinking coupled with a pragmatism that does not reveal checklists, per se, but provides some of their ultimate function. Jotted notes from these expositions will likely provide something akin to that elusive checklist, designed to most efficiently win wars or, even better, to prevent them.

One will also discover that this quest for an understanding of naval strategy is more than a source of professional development. It is that to be sure, but there is also an existential need for a broader understanding of why a navy is so vital to a maritime nation such as ours, not only to *utilize* naval power but to *preserve* it as well.

This is evident in the realization that the so-called maritime strategy, which is periodically devised or revised by the chief of naval operations and his contemporaries in the Marine Corps and Coast Guard, serves not only as a guiding document for professionals but a form of "advertisement" for the sea services as well.

At a "Maritime Security Dialogue" conference in October 2014, sponsored by the Center for Strategic and International Studies, Chief of Naval Operations Jonathan Greenert included in his opening remarks the admonition that "presence is a priority." Coincidentally, a short article titled "Presence" appeared at the same time in the November 2014 *Proceedings* as a "Lest We Forget" column. As self-serving as it may seem, I begin this collection with that article as

Chapter 1, not to point out that "great minds think alike," but because I wrote it for the very same reasons that Admiral Greenert places such emphasis on this core capability of the U.S. Navy. We were both preaching from the naval pulpit because, for all their importance, the primary missions—the raison d'être—of the Navy (forward presence, all-domain access, deterrence, sea control, power projection, and maritime security), are not intuitively obvious.

One reason for this is that each of these missions requires at least some explanation to place them in their proper (maritime) context. As indicated in Chapter 1, the concept of forward presence was not understood nor appreciated by the young ensign who was a cog in the machine that was carrying it out. The problem is exacerbated when extended to average American citizens (and, increasingly, to members of Congress, from whence they come) who are ultimately paying for these expensive naval forces.

Another less obvious reason for the underappreciation for the importance of the Navy is that it suffers from an image problem that is ironically caused to some degree by its incredible success in the middle of the last century—something that Samuel Huntington addresses in Chapter 3 of this anthology.

The unquestionable glory of the Pacific War (1941–45), where American Sailors fought and won many awe-inspiring battles—some of them gargantuan in scope—elevated the American Navy's status in the eyes of its citizens to unprecedented heights. And while that elevation was richly deserved and should never be forgotten, it had the converse effect of setting a very high bar for future naval operations in terms of public perception. Except for a minority of unusually informed citizens, most Americans perceive their Navy as expensive (true) and not as necessary as it was in those harrowing days of World War II (decidedly not true).

One can look to the battles of Midway, Leyte Gulf, and Okinawa and see the tangible results of sunken ships, captured islands, and an ultimate surrender ceremony on the deck of a battleship. One must rely on much less tangible "proofs" when trying to visualize such things as "forward presence" or "sea control."

The Navy's role in the Jordanian Crisis described in Chapter 1 is but one of countless examples of naval power being wielded in ways that do not make good subjects for action movies yet are essential to the nation's ability to maintain its security and to influence world affairs. Further, few shoppers at Walmart ever stop to consider that the plethora of products they find on those rows and rows of shelves are there partly because the Navy is doing its job, keeping the sea lanes of trade open 24/7/365. In Chapter 7, Norman Friedman addresses this problem: "That we depend on imports means that we have vital interests in far corners of the world. It happens that relatively few Americans understand as much."

Dr. Friedman also points out that even though World War I was largely a maritime war, that aspect is overshadowed by images of trench warfare. The same problem exists for the Korean War, which does not even occur (at least the parts played by the United States and the United Nations) without American dominance in the realm of sea control. The absence of great sea battles, such as had taken place just a few years before, obscures that relevant truth, leaving a popular image of the Korean conflict as an Asian land war fought only by Soldiers and Marines with only a slight nod to naval air power. Similarly, the popular understanding of the outcome of the harrowing Cuban Missile Crisis of 1962 acknowledges the naval "quarantine" but often fails to appreciate the essential role of credible deterrence that kept events from spinning out of control.

So an appreciation of naval strategy is important not only to naval professionals who must plan and fight the nation's wars, but also to those individuals who must convince their fellow citizens to invest a substantial portion of their national wealth in something that they can rarely see and often do not understand. Those citizens (taxpayers) and their legislative representatives hold the keys to the nation's security, and it is imperative that the Navy be able to convey the reasons for its existence through an explication of what it does and must be prepared to do. A study of naval strategy provides the framework for that vital discussion.

While this collection of Naval Institute writings on strategy is not likely to be read by many "average Americans," those professionals who do read these informative and sometimes provocative pieces will likely be reminded of the complexity of the subject and will be better equipped to deal with it not only as a part of their profession but also as a means to spread the "gospel" of why we, as a maritime nation, need a navy.

Here are a chosen few of the many articles and books that the Naval Institute has contributed to this needed "polylogue" among academic thinkers, naval professionals, and the American citizenry. In the pages that follow, the mysteries and verities of naval strategy are explored, explained, and debated to better equip those who must devise strategy, those who must carry it out, and those who need to have a basic understanding of why we maintain a navy in the first place.

1 "PRESENCE"

LCDR Thomas J. Cutler, USN (Ret.)

As explained in the Introduction, this short article focuses on "forward presence," one of the core capabilities described in the sea services' joint maritime strategy, *A Cooperative Strategy for 21st Century Seapower.*

"PRESENCE"

By LCDR Thomas J. Cutler, USN (Ret.), U.S. Naval Institute *Proceedings* (November 2014): 93.

The following is an entry in a personal journal—called "The Log" by its author—kept by a young ensign serving as Assistant G-Division Officer in the carrier *Independence*, deployed to the Mediterranean in the fall of 1970:

> 21 SEP – Rumors are flying; things are happening. JFK is on her way to join Sara[toga] and us. King Hussein has in no uncertain terms asked for American air strikes immediately as well as subsequent troop intervention—this stemming from the fact that Syria has entered the conflict with an armored invasion. G DIV had a major load again—this time we loaded anti-tank weapons. (THE PLOT THICKENS).

> The ordnance is at this moment (2130) hanging on the aircraft (mostly A-4s with some A-6s). I have a strong feeling that an air strike is coming soon.

He was wrong. No airstrike was launched, and G-Division subsequently set about removing the anti-tank weapons from the aircraft and returning them to the ship's magazines many decks below. As is often the case with young people who have not yet come face-to-face with the realities of war, the ensign recorded his disappointment, lamenting that "we are only going to be a *presence* in this situation and are not going to *do* anything."

But what this young officer did not fully appreciate was that being a "presence" was of immeasurable significance, that this giant warship with such immense potential power had played a role on the world stage with ramifications that resonated not only in the troubled Middle East but in the halls of the Kremlin and the White House. What he viewed as anticlimax was instead the achievement of strategic objectives in a manner that would have pleased Sun Tzu, the ancient philosopher of war who famously said, "The supreme art of war is to subdue the enemy without fighting."

While nodding off during sea power lectures at officer candidate school, this neophyte had apparently missed the part explaining that an effective navy does so much more than engage in combat. He apparently was awake during the part about projecting power ashore but failed to appreciate the importance of the Navy's other vital missions—much less tangible, but every bit as important—of sea control, deterrence, and forward *presence*.

What had occurred those many years ago is what is known to historians as "The Jordanian Crisis of 1970," a complicated situation that involved Iraq, Syria, the Palestinian Fedayeen, Israel, the United States, the Soviet Union, a terrorist organization known as "Black September," and, of course, Jordan. At the height of the crisis, the Jordanian Army was able to repulse an armored incursion by Syria, something that would have been seriously jeopardized had the Soviets intervened on the side of their Syrian ally.

The latter was prevented by a strong rhetorical stance by the American president made credible by a great deal of potential power in the form of the

U.S. Sixth Fleet providing that "presence" noted in the ensign's "Log." Several attack aircraft carriers, a strong amphibious element, and many more units of the Sixth Fleet positioned in the eastern Mediterranean provided a show of force that allowed the U.S. to achieve its regional objectives while preventing this highly volatile situation from exploding into something much worse. An important element in the outcome of this delicate situation was the time factor, because this persuasive force had been accumulated much more rapidly than other elements of U.S. or Soviet power could be brought to bear, proving the importance not only of presence but of *forward* presence, something a deployed navy provides so effectively.

In the end, an important Middle Eastern ally had been preserved while keeping the USSR at bay. As the crisis abated, "The Log" turned to other subjects of great importance, such as assessing the next liberty port for the possibility of conquests of a different sort.

2 "MISSIONS OF THE U.S. NAVY"

VADM Stansfield Turner, USN

In this foundational article, Admiral Turner examines the "quartet of missions"—strategic deterrence, sea control, projection of power ashore, and naval presence—that have become the basis of the Navy's current maritime strategy. His analysis includes a discussion of how these missions evolved and an assessment of their worth, as well as various associated tactics that can be applied.

"MISSIONS OF THE U.S. NAVY"

By VADM Stansfield Turner, USN, U.S. Naval Institute *Proceedings* (December 1974): 19–25.

A quartet of missions has evolved—Strategic Deterrence, Sea Control, Projection of Power Ashore, and Naval Presence. Inevitably, there are overlaps and interdependence, but we can understand the Navy better if we assess each mission individually. We must know what each mission's objectives are so that we do not overlook some useful new tactic or weapon and so that we can strike the proper balance whenever these missions compete for resources.

One of the important challenges facing naval officers today is to define, then articulate, why we need a navy and what it should be able to accomplish for the country. The changes in national attitudes and military technology and the relationship of nations today are such that we cannot accept as sacrosanct the traditional rationale for a navy. We must reexamine and be willing to change the well-established missions of our Navy. In 1970 the Chief of Naval Operations defined the current missions of the U.S. Navy as being Strategic Deterrence, Sea Control, Projection of Power Ashore, and Naval Presence. By "missions" he meant the outputs or objectives of having a navy. As a starting point, we should examine how these four missions evolved. We can then ask what they specifically mean today and whether they are an adequate rationale for a navy.

The first mission of the earliest navies was being able to move military forces by sea. As time went on, there were many technological milestones, new tactical concepts, and maritime initiatives, but the basic naval mission to ensure the safe movement of ground forces and their supplies across the sea endured for centuries.

By the 18th century, however, sea trade routes were flourishing, exploration was becoming more far ranging, the horizons of imperialism were widening, commerce was growing, and with it, piracy. As nations began to depend on the seas for their economic well being, they needed security of movement by sea. Control of the sea became the *sine qua non* of economic growth. The Sea Control mission expanded to include the protection of shipping for the nation's economy as well. At the end of the 19th century, Alfred Thayer Mahan defined maritime power to include merchant marine and naval forces plus all of the bases and coaling stations needed to support each. He popularized the "control of the seas" concept as a key to expanding national power and prestige. To Mahan the term "control of the seas" meant both denying the enemy use of the seas and asserting one's own use, both with a battle fleet superior to that of the enemy.

British and German naval strategy in World War I reflected this heritage. Both navies believed that a decisive battle fleet encounter would determine control of the seas. Hence, tactical caution dominated the Battle of Jutland. After that failure to defeat the British battle fleet, the Germans challenged British seapower indirectly, first with surface commerce raiders, and later with unrestricted submarine warfare. The British reacted by attempting to blockade the German U-boats with mines laid across the exits from the North Sea. This failed, and the World War I struggle for control of the Atlantic evolved into a grueling war of attrition. Large numbers of allied antisubmarine ships and aircraft were pitted against a much smaller number of German submarines. Despite the difference between this kind of warfare and the classic concept that battle fleet engagements would determine control of the seas, few strategists understood how radically the concept of "control of the seas" had been altered by the advent of the submarine. British, German, Japanese, and American preparations for World War II all concentrated on potential battle fleet actions. Only a few voices pointed out that an additional submarine might be more useful than another battleship.

Equally few strategists forecast the dominant role that control of the air over a surface fleet would have. However, in March 1941, off Cape Matapan in Greece, the first engagement of major surface forces since Jutland demonstrated that it was the presence of a British aircraft carrier that allowed an otherwise weaker force to prevail. Throughout World War II the primary use of naval carrier-based air power was in the sea control role of defeating enemy carriers and battleships, with a secondary role of providing close air support for amphibious assaults. By the end of World War II the idea of totally denying the seas to one's enemy while asserting one's own exclusive use had been overtaken by technology. On the one hand it was nearly impossible to deny an enemy submarine fleet access to the seas; on the other, there were likely to be areas of the sea where enemy air power would make the assertion of one's presence prohibitively costly. Yet, for the first several decades after the second World War, the U.S. Navy had such a monopoly on seapower that the term "control of the seas" understandably continued to carry its long-established connotation.

The new term "Sea Control" is intended to acknowledge the limitations on control of the oceans brought about by the advent of the airplane and the submarine. It connotes a more realistic concept of control in limited areas and for limited periods of time. It is conceivable for a navy today to temporarily exert air, surface, and subsurface control in a limited area while moving ships into position to project power ashore or to resupply overseas forces. But it is not conceivable, except in the most restricted sense, to totally control the seas for one's own use, or to totally deny them to an enemy.

This may change with evolving technology and tactics but, in the meantime, we must approach the use of the term "Sea Control," from two directions: *denying* an enemy the right to use some seas at some times; and, *asserting* our own right to use some seas at some times. Any seapower may both assert its own use of the seas and deny that right to the enemy at any given time. Its efforts will usually be divided between the two objectives. For instance, if the United States were attempting in wartime to use the North Atlantic to reinforce Europe, it would put the greater percentage of its effort on asserting sea control. In a situation like the war in Vietnam, where our use of the seas was not challenged, we made a substantial effort to deny the other side any seaborne infiltration into South Vietnam.

There are four different tactical approaches for achieving these Sea Control objectives:

Sortie Control. Bottling up an opponent in his ports or his bases is a most economical means of cutting off a nation's use of the seas or ability to interfere.

Nevertheless, no blockade is 100% successful. Some units may be beyond the blockade when hostilities commence and will remain to haunt opposition forces. Against the enemy's aircraft there is no static defense. Planes must be bombed at their bases. If we assume an opponent will be in control of the air near his ports, sortie control tactics must primarily depend on submarines and mines. Thus we must conclude that blockades are weapons of attrition requiring time to be effective. But the lesson of history is perhaps the most instructive of all—ingenious man has usually found ways to circumvent blockades.

Choke Point Control. Sometimes the best place to engage the enemy is in a geographical bottleneck through which he must pass. In so doing, platforms like ASW aircraft that probably could not survive in the area of the enemy's sortie point can be used.

Open Area Operations. Once enemy ships, submarines, or aircraft are loose in, or above, the open ocean, we have the option of instituting search procedures. Open area search is a third form of attrition operations. In short, these operations are not part of defending specific merchant or naval shipping. They are intended to seek out the enemy and reduce the threat before it makes contact with forces to be protected.

Local Defense. In contrast to searching out a large area with the intent of locating, tracking and possibly prosecuting and destroying enemy forces, in local defense we assert our use of the seas. If our attrition forces have not been 100% successful, the enemy may be able to close our forces to within range of attack. If so we must defend ourselves by (1) engaging his attack platforms directly, (2) defeating his attack weapons by direct kill, or (3) decoying or deceiving his weapons. This objective may include contributions, as a preliminary, of sortie control, choke point control and open area operations. Depending upon our purpose in asserting use of the seas we may select (1) to try to evade and deceive, (2) to close in and attack, and (3) to attack enemy forces when they close to within their weapon range, and to defend actively or passively against these weapons.

In executing Sea Control, our relative emphasis on these four tactics will vary with the particular circumstance, especially the enemy threat and our own objectives. For instance, if our objective is to ensure an early flow of supplies to some theater of war, attrition type tactics may not be adequate. Or, if in the early days of a conflict, the enemy has dense defenses near his ports and bases, sortie control may be difficult.

Additionally, in executing Sea Control tactics, two passive techniques deserve particular mention:

Deception. Assertive Sea Control objectives do not necessarily demand destruction of the enemy's force. If the enemy can be sufficiently deceived to frustrate his ability to press an attack, we will have achieved our Sea Control objective. Force routing, deceptive/ imitative devices, and other anti-search techniques can be employed, often in combination with other tactics.

Intimidation. The perceptions of other nations of our Sea Control capability relative to that of other major powers can influence military decisions. What a nation says about its capabilities can influence the challenges that are offered or accepted. A Sea Control force that is recognized by the enemy may inhibit the enemy's willingness to commit his sea denial forces.

By the early 19th century, another important naval mission had evolved—the projection of ground forces from the sea onto the land. Modern amphibious warfare began during the wars of the French Revolution. A new dimension in tactics was given to commanders in the Projection of Power Ashore through amphibious assault. During World War I the first major amphibious assault of the 20th century was attempted at Gallipoli. The failure of the assault as a result of poor execution nearly killed the amphibious assault concept. In World War II, however, amphibious assault played a major role in both theaters, and the Inchon assault in Korea in 1950 constituted a stunning tactical maneuver.

Amphibious assault tactics are largely a function of the size of the operation. The war in Korea and later the war in Vietnam brought into play two new ways of projecting power ashore: naval bombardment and naval tactical air. Naval bombardment was undoubtedly used on occasions as far back as the 18th century to interfere with enemy coastal communications and installations. But, by 1950, it was employed primarily as a part of amphibious assault. Both Korea and Vietnam have long, exposed coastlines with significant road and rail lines.

Here naval bombardment came into its own as an independent way of projecting power ashore.

Naval bombardment is presently available from naval guns in destroyers utilizing two tactics: direct fire and indirect fire. If the target is visually observable from the firing ship, direct fire is the simplest and most accurate method. If it is not, fire directed by a spotter on the beach, from an aircraft or by prearrangement based on geographical coordinates, must be employed. During World War II, naval tactical air began moving into the projection of power role. Tactical air projection evolved fully as a mission in the post–World War II period. The marriage of the jet aircraft and improved, lesser drag munitions gave the aircraft carrier a capability of extending its reach far past the shoreline.

During the Korean War, naval tactical air came to play a major role in support of the land campaign: air attacks on enemy networks; transportation; air superiority over the battlefield; and close air support of ground forces.

The four basic tactical air tactics are: deep interdiction; battlefield interdiction; close air support; and counter-air/anti-air warfare.

Deep Interdiction is usually more strategic than tactical. That is, the impact on the ground campaign is more remote and less immediately felt than in the other tactical air tactics. Attacks can be either directed at the enemy's war making potential, that is against civilian morale, economy, or command structure; or they can be militarily disruptive, that is against military bases, logistics sources, depots, or supply routes. Generally these are fixed targets that are pre-planned. Thus such attacks may include advance preparations of target lists, strike group tactics, approach profiles, defense suppression techniques, and planned staging of attacks. For instance, attack aircraft if threatened by enemy fighters, have three options: continue to target, jettison ordnance and attempt to escape or, in some circumstances, jettison and attempt to engage the fighters. The actual choice will depend on the type of aircraft and the nature of the mission flown. One or another of these tactics may be preferred, but it generally will be determined in advance.

In addition to the primary attack aircraft, a typical operation may also involve special EW, anti-ground defense, or air-to-air configured fighters (see counter-air tactic below), which may be preceded by photo electronic reconnaissance missions. Because deep interdiction operations are remote from the fluid conditions in battle areas, enemy air defenses generally tend to be well integrated. Reasonable radar warning, good air control facilities, airborne patrol or ground alert fighters, and both surface-to-air missiles (SAMs) and anti-aircraft fire (AAA) can be anticipated.

Battlefield Interdiction is generally carried out in the enemy's division to corps area, about 5 to 50 kilometers into enemy-held territory from the forward edge of the battle area (FEBA). It is directed primarily against military targets, both static, such as bridges, gun emplacements, and bunkers, and fleeting, such as troop concentrations or vehicular traffic. The measure of effectiveness of this tactic may not be so much in targets destroyed as in the denial to the enemy of ground force mobility behind the front lines or of increased difficulty in resupplying key items such as POL and ammunition.

Flights and areas of coverage are usually pre-planned but include targets of opportunity with target acquisition being the mission-limiting feature, especially in bad weather, at night, or in jungle terrain. Real-time intelligence may be provided by various types of sensors. In most circumstances, attacking aircraft will penetrate enemy-held territory over short distances only. Because of this, they may well operate at low levels with or without fighter cover. Mobile SAMs, although perhaps not well netted and with only short radar warning time available, plus AAA, can be anticipated. There must be good air space control and coordination because of the close proximity to front line close air support operations.

Close Air Support operations are "call fire" in response to direct requests from ground units or through Forward Air Controllers (FACs). As these same sources may be calling simultaneously for artillery, good liaison procedures between the controllers and the close support aircraft are essential. Aircraft, capable of heavy ordnance loads and low level operations, must be able to scramble quickly from nearby bases or be able to loiter in the area on call. The zone of operations tends to be no deeper than five kilometers from friendly forces, hence accuracy of weapon delivery is extremely important. While large, mobile SAMs are possible in this zone, they are less a threat than AAA and hand held, short-range SAMs. Fighter escort is usually not assigned, though there may be general front coverage against raiding aircraft.

Counter-air/Anti-air. There are two distinct air superiority operations:

- Counter-air to neutralize an enemy's anti-air capabilities sufficiently to minimize attrition to our attacking forces.
- Anti-air operations to deny an enemy the capability of operating attack aircraft in our areas of interest and control.

Escorting fighter aircraft provide counter-air defense against enemy fighters; armed suppression fighters or attack aircraft are directed against ground defenses such as SAMs and their control facilities; and a variety of EW aircraft techniques defend against SAMs and AAA. Escort fighter aircraft tactics can be divided into:

- Long range intercept utilizing sophisticated radar and fire control techniques and air-to-air missiles.
- Dog-fight maneuvering and close in missiles or guns.

In the long range case, positive identification is a critical problem. In either case, air intercept control can be valuable.

For anti-air operations, airborne CAP, ground alert fighters, SAMs, AAA and deceptive electronic measures can be employed.

This type of mission tactic categorization can be valuable in assessing own and enemy capabilities, whereas the usefulness of a simple summary of weapon/ aircraft performance characteristics is limited. When we thereupon superimpose weapons and aircraft, we can then better evaluate our capabilities for achieving some part of the tactical air projection mission. By superimposing enemy defensive capabilities, we can move into a dynamic evaluation of our systems in combat. By superimposing enemy offensive weapons and aircraft and our defensive systems we can estimate our vulnerabilities.

Beyond this, we can also use this categorization to identify the weapons and aircraft characteristics that we need for each tactic. There will be specific scenarios where some of the judgmental evaluations will be incorrect. It would be desirable to be infinitely flexible and have maximum characteristics in all aircraft and weapons. Unfortunately, the laws of both physics and economics prevent that. Hence, some evaluation of probable use and likely need can be valuable.

Only a fine distinction separates some aspects of the Sea Control and Projection of Power Ashore missions. Many weapons and platforms are used in both missions. Amphibious assaults on choke points or tactical air strikes on enemy air bases can be employed as a part of the Sea Control mission. Sea-based tactical aircraft are used in Sea Control missions for anti-air warfare and against enemy surface combatants. The distinction in these cases is not in the type of forces nor the tactics which are employed, but in the purpose of the operation. Is the objective to ensure/prevent use of the seas or is it to directly support the land campaign? For instance, much of the layman's confusion over aircraft carriers stems from the impression that they are employed exclusively in the Projection of Power Ashore role. Actually, from the Battle of Cape Matapan through World War II, aircraft carriers were used almost exclusively to establish control of the ocean's surface. Today they clearly have a vital role to play in both the Sea Control and Projection of Power missions.

Both Sea Control and Projection of Power Ashore can be termed "war fighting" missions. We buy forces capable of executing these missions in combat. The Navy's two other missions are "deterrent." We buy forces to ensure against having to engage in combat.

During the 19th century, the term "gunboat diplomacy" came into the naval vocabulary. In the quest for colonies, powerful nations paraded their naval forces to intimidate and serve warning. In time, the range of this activity extended from warnings and coercion to demonstrations of good will and humanitarian assistance. Today, the Naval Presence mission is the use of naval forces, short of war, to achieve political objectives.

We attempt to accomplish these objectives with two tactics: *preventive deployment* and *reactive deployment.* A preventive deployment is a show of presence in peacetime whereas a reactive deployment is a response to a crisis. In a preventive deployment, force capabilities should be relevant to the problem, clearly not inferior to some other naval force in the neighborhood, and in capability we should have a reasonable hope that reinforcements can be made available if necessary. On the other hand, a reactive deployment may or may not actually involve a movement or deployment of forces. There will be instances when the threat of doing so, perhaps communicated through an alert or mobilization order, will produce a desired reaction in itself. When a force is deployed, however, it needs to possess an immediately credible threat and be prepared for any contingency. A comparison with other naval forces in the area will be inevitable.

In deciding to insert a presence force, the size and composition of force must be appropriate to the situation. There are five basic actions which a Naval

Presence force can threaten: *amphibious assault, air attack, bombardment, blockade,* and *exposure through reconnaissance.*

Almost any size and type of presence force can also imply that the United States is concerned with the situation and may decide to bring other military forces or non-military pressures to bear. All too often, especially in reactive deployments, we tend to send the largest and most powerful force that can move to the scene rapidly. The image created may not be appropriate to the specific problem.

When selecting a *Naval Presence* force, we must also take into account how the countries that we want to influence will perceive the situation. There are three distinctly different categories of national perceivers:

The Soviet Union. When contemplating a U.S. presence force, the Soviets must assess their comparative naval strength available over time, and the expected degree of U.S. resolve. As the United States is not likely to threaten the U.S.S.R. directly, except in a world-wide crisis of the most serious proportions, the principal strength comparison would probably be on which country could actually exercise sea control in the area in question.

Nations Allied to the Soviets. Nations with close ties to the Soviets must assess relative U.S.–U.S.S.R. capabilities. These powers will ask, "can the United States project its assembled power onto my shores?" and "can the U.S.S.R. deny them that capability." Thus, third nation appraisal of relative sea control strengths may be the most critical factor and their assessment may not correspond to either U.S. or Soviet assessments of identical military factors.

Unaligned Third Nations. There will be cases where a nation is not able to invoke major power support in a dispute with the United States. The perceptions of such a country would likely focus on U.S. capability and will to project its power ashore to influence events in that country itself.

Thus, the Naval Presence mission is sophisticated and sensitive and, because of the subtleties involved, probably the least understood of all Navy missions. A well orchestrated Naval Presence can be enormously useful in complementing diplomatic actions. Applied deftly but firmly, in precisely the proper force, Naval Presence can be a persuasive deterrent to war. Used ineptly, it could be disastrous. In determining presence objectives, scaling forces, and appraising perceptions, the human intellect must take precedence over ships and weapons systems.

The second naval deterrent mission came with the introduction of Strategic Deterrence as a national military requirement. Again, the combination of improved aircraft performance and smaller packaging of nuclear weapons made the aircraft carrier capable of contributing to this new mission. With the Navy struggling to readjust its missions to peacetime needs and with the U.S. Air Force at that time establishing its own place in the military family, it is understandable that there was a sense of competition for this new role. However, by the mid-1960s the development of the Polaris submarine eliminated any question of appropriateness of this mission for the Navy.

Our Strategic Deterrence objectives are:

- to deter all-out attack on the United States or its allies;
- to face any potential aggressor contemplating less than all-out attack with unacceptable costs; and to maintain a stable political environment within which the threat of aggression or coercion against the United States or its allies is minimized.

In support of these national objectives, we have three principal military "tactics" or force preparedness objectives. The first is to maintain an *assured second strike* capability in the hope of deterring an all-out strategic nuclear attack on the United States. Today that means dissuading the Soviets from starting a nuclear war. We hope to achieve this by maintaining a strategic attack force capable of inflicting unacceptable damage on any enemy even after he has attacked us. The Navy's Polaris/Poseidon/Trident forces are fundamental to this deterrence because of their high nuclear survival probability.

A second tactic is to design our forces to ensure that the United States is not placed in an unacceptable position by a partial nuclear attack. If the Soviets attacked only a portion of our strategic forces, would it then make sense for the United States to retaliate by striking Soviet cities, knowing that the Soviets still possessed adequate forces to strike our cities? This means making our strategic strike forces quickly responsive to change in targeting and capable of accurate delivery. SSBN forces can be well tailored to these requirements.

A third objective is to *deter third powers* from attacking the United States with nuclear weapons. Because of the great disparity between any third country's nuclear arsenal and ours, the same forces deterring the Soviet Union should deter others.

Finally, we maintain sufficient strategic forces so that we do not appear to be at a disadvantage to the Soviet Union or any other power. If we were to allow the opinion to develop that the Soviet strategic position is markedly superior to ours we would find that political decisions were being adversely influenced. Thus we must always keep in mind the *balance of power image* that our forces portray to the non-Soviet world. In part, this image affects what and how much we buy for strategic deterrence. In part, it affects how we talk about our comparative strength and how we criticize ourselves.

In summary, the Strategic Deterrence mission is divided into four tactics: *Assured Second Strike, Controlled Response, Deterrence of Third Powers,* and *Balance of Power Image.*

There is very little overlap between Strategic Deterrence and other Navy mission areas at present. However, significant improvements in enemy ASW technology could reduce the ability of SSBNs to survive without assistance from friendly Sea Control forces. With this exception and the fact that aircraft carriers still possess the potential for nuclear strikes, that mission is performed almost exclusively by forces designed specifically for it.

There is a good deal of overlap in the overall field of deterrence. There is no doubt that our strategic deterrent forces inhibit at least ourselves and the Soviet Union from engaging in non-strategic or conventional warfare. It is also true that the very existence of our Sea Control and Projection of Power Ashore forces deter conventional warfare, over and above whether we consciously employ them in the Naval Presence role. There is very likely even some interplay between our conventional force capability and the way in which our strategic deterrent forces are perceived, e.g., a Sea Control capability is essential to the security of our sea-based strategic deterrent forces. Thus, the boundary lines between the four naval mission areas cannot be precise. More than anything, they each express a somewhat different purpose. Despite these inevitable overlaps and interdependence, we can understand the Navy far better if we carefully examine each mission individually. We must know what each mission's objectives are so that we do not overlook some useful new tactic or weapon and so that we can strike the proper balance whenever these missions compete for resources.

Additionally, we must be careful not to view as rigidly fixed these mission areas and their relative importance. We swung from a primary emphasis on Sea Control with a secondary interest in amphibious assault before and during World War II, to a primary emphasis on strategic deterrence and tactical air projection for the 20 to 25 years following that war. In about the mid-1960s, the dramatic and determined growth of Soviet naval capabilities forced renewed attention to Sea Control. The even more recent national disinclination to engage ground forces in support of allies should perhaps today place more attention on the conventional deterrent mission of Naval Presence. The dynamic nature of world conditions will demand a continuing reassessment of the relation of one mission to another and the comparative emphasis on their individual tactics.

Perhaps this constant flow and counter flow of mission emphasis and tactical adaptation is even more accentuated today than in the past. On the one hand, the pace of technological innovation is forcing this. On the other, the changing nature of world political relationships and domestic attitudes demands a continual updating of naval capabilities to support national policy. Naval officers, as professionals, must understand the Navy's missions, continually question their rationale, and provide the intellectual basis for keeping them relevant and responsive to the nation's needs. Unless we do, we will be left behind, attempting to use yesterday's tools to achieve today's objectives.

3 "NATIONAL POLICY AND THE TRANSOCEANIC NAVY"

Samuel P. Huntington

Long before Samuel Huntington wrote his famous 1993 *Foreign Affairs* article "The Clash of Civilizations?," which he expanded into a book (*The Clash of Civilizations and the Remaking of World Order*) in 1996, Professor Huntington wrote this thought-provoking article in *Proceedings* magazine. Those versed in classical strategic writings will recognize Clausewitz in his opening sentence but further reading will reveal that he is at odds with Mahan, essentially contending that the iconic maritime strategist is no longer relevant in light of the changing world situation.

Despite his Cold War focus, much of what he offers remains worthy of consideration if one substitutes "our enemies" where references are made to the Soviet Union. Despite a different world order from that of the 1950s when Huntington wrote this article, much of what he proffers is potentially relevant. He was amazingly prescient in his views about the littoral and projection of power from the sea, and his recognition of the importance of public perception as to the role (and importance) of a Navy rings true today.

"NATIONAL POLICY AND THE TRANSOCEANIC NAVY"

By Samuel P. Huntington, U.S. Naval Institute *Proceedings* (May 1954): 483–93.

I

The Elements of a Military Service

The fundamental element of a military service is its purpose or role in implementing national policy. The statement of this role may be called the *strategic concept* of the service. Basically, this concept is a description of how, when, and where the military service expects to protect the nation against some threat to its security. If a military service does not possess such a concept, it becomes purposeless, it wallows about amid a variety of conflicting and confusing goals, and ultimately it suffers both physical and moral degeneration. A military service may at times, of course, perform functions unrelated to external security, such as internal policing, disaster relief, and citizenship training. These are, however, subordinate and collateral responsibilities. A military service does not exist to perform these functions; rather it performs these functions because it has already been called into existence to meet some threat to the national security. A service is many things: it is men, weapons, bases, equipment, traditions, and organization. But none of these have meaning or usefulness unless there is a unifying purpose which shapes and directs their relations and activities towards the achievement of some goal of national policy.

A second element of a military service is the resources, human and material, which are required to implement its strategic concept. To secure these resources it is necessary for society to forego the alternative uses to which these resources might be put and to acquiesce in their allocation to the military service. Thus, the resources which a service is able to obtain in a democratic society are a function of the *public support* of that service. The service has the responsibility to develop this necessary support, and it can only do this if it possesses a strategic concept which clearly formulates its relationship to the national security. Hence this second element of public support is, in the long run, dependent upon the strategic concept of the service. If a service does not possess a well defined strategic concept, the public and the political leaders will be confused as to the role of the service, uncertain as to the necessity of its existence, and apathetic or hostile to the claims made by the service upon the resources of society.

Organizational structure is the third element of a military service. For given these first two elements, it becomes necessary to group the resources allocated by society in such a manner as most effectively to implement the strategic concept. Thus the nature of the organization likewise is dependent upon the nature of the strategic concept. Hence there is no such thing as the ideal form of military organization. The type of organization which may be appropriate for one military service carrying out one particular strategic concept may be quite inappropriate for another service with a different concept. This is true not only in the lower realms of tactical organization but also in the higher reaches of administrative and departmental structure.

In summary, then, a military service may be viewed as consisting of a strategic concept which defines the role of the service in national policy, public support which furnishes it with the resources to perform this role, and organizational structure which groups the resources so as to implement most effectively the strategic concept.

Shifts in the international balance of power will inevitably bring about changes in the principal threats to the security of any given nation. These must be met by shifts in national policy and corresponding changes in service strategic concepts. A military service capable of meeting one threat to the national security loses its reason for existence when that threat weakens or disappears. If the service is to continue to exist, it must develop a new strategic concept related to some other security threat. As its strategic role changes, it may likewise be necessary for the service to expand, contract, or alter its sources of public support and also to revamp its organizational structure in the light of this changing mission.

II

The Crisis of the Navy

That the United States Navy was faced with a major crisis at the end of World War II is a proposition which will hardly be denied. It is not as certain, however, that the real nature and extent of this crisis has been so generally understood. For this was not basically a crisis of personnel, leadership, organization, material, technology, or weapons. It was instead of a much more profound nature. It went to the depths of the Navy's being and involved its fundamental strategic concept. It was thus a crisis which confronted the Navy with the ultimate question: What function do you perform which obligates society to assume responsibility for your maintenance? The crisis existed because the Navy's accustomed answer to this question—the strategic concept which the Navy had been expressing and the public had been accepting for well over half a century—was no longer meaningful to the Navy nor convincing to the public.

The existence of this crisis was dramatically symbolized by the paradoxical situation in which the Navy found itself in 1945: It possessed the largest fleet in its history and superficially it had less reason to maintain such a fleet than ever before. The fifteen battleships, one hundred aircraft carriers, seventy cruisers, three hundred and fifty destroyers, and two hundred submarines of the United States Navy floated in virtually solitary splendor upon the waters of the earth. It appeared impossible, if not ridiculous, for the Navy still to claim the title of the Nation's "first line of defense" when there was nothing for the Navy to defend the nation against.

Critics of the Navy were not slow in undermining the latter's public support by pointing out these paradoxes. As one high ranking Air Force officer put it:

> Why should we have a Navy at all? The Russians have little or no Navy, the Japanese Navy has been sunk, the navies of the rest of the world are negligible, the Germans never did have much of a Navy. The point I am getting at is, who is this big Navy being planned to fight? There are no enemies for it to fight except apparently the Army Air Force. In this day and age to talk of fighting the next war on the oceans is a ridiculous assumption. The only reason for us to have a Navy is just because someone else has a Navy and we certainly do not need to waste money on that.

The public appeal of this simple logic was enhanced by the widespread postwar reaction against the military, the popular desire to reduce the defense budget, and the fact that one of the Navy's sister services possessed in intercontinental atomic bombing a strategic concept which seemed to promise a maximum of security at a minimum of cost and troublesome intervention in world politics. It is hardly surprising that as a result a 1949 Gallup Poll revealed that 76% of the American people thought that the Air Force would play the most important role in winning any future war whereas only 4% assigned this role to the Navy.

This lack of purpose had its organizational implications also. Most important among these was the tendency to increase naval opposition to unification of the armed forces. Without an accepted strategic concept the Navy had to rely upon organizational autonomy rather than uniqueness of mission to maintain its identity and integrity. This had additional unfortunate implications for naval public support, however, since it enabled its critics to paint the picture of a willful group of die-hard admirals opposing unification for purely selfish purposes.

The causes of this crisis of purpose and its unfortunate political and organizational implications were to be found, of course, in the redistribution of international power which occurred during World War II, the new threats to American national security which emerged after the War, and the consequent shifts in American foreign policy to meet these threats. The critics of the Navy argued in effect that these changes left the Navy without a strategic concept relevant to the postwar world. If they were to be proved wrong and if the Navy were not to be reduced to a secondary service concerned exclusively with protection of supply lines, the Navy must find a new role for itself in national policy. It is the principal thesis of this article that out of the postwar uncertainty, demoralization, and confusion, there has developed a new naval doctrine which realistically relates the Navy to national goals. The substance of this concept has already been described and formulated by a number of naval writers and leaders, and the development of this doctrine must eventually have a significant effect on the public support and organization of the Navy. This doctrine, however, will require a fundamental revolution in naval thinking. Consequently before describing it in detail, it will be appropriate to consider briefly the nature of the relation between the Navy and national policy in the past.

III
The Navy and National Policy: Continental Phase

The first stage of American national security policy may best be described as the Continental Phase. This lasted approximately from the founding of the Republic down to the 1890s. During this period the threats to the national security arose primarily upon this continent and were met and disposed of on this continent. The limited capabilities of the United States during these years did not permit it to project its power beyond the Western Hemisphere. And, indeed, the history of this period may also be interpreted as the history of the gradual struggle by the United States for supremacy within the American continent. This policy manifested itself in our refusal to enter into entangling alliances with non-American powers, in our promulgation and defense of the principles of the Monroe Doctrine, and in our gradual expansion westward to the Pacific.

During these years those threats which arose to the national security were generally dealt with on land, and sea power consequently played a subordinate role in the implementation of national policy. The most persistent security threat, of course, came from the Indian tribes along the western and southern frontiers. These could only be met by the army and the militia. Similarly during the War of 1812 the American Navy was unable to prevent the British from reinforcing Canada, seizing and burning the national capitol, and landing an army at New Orleans. Instead, each of these threats had to be countered by what land forces there were available. The Mexican War was likewise primarily an army affair, although the Navy in the closing campaign of the war performed yeoman service in landing Scott's army at Vera Cruz. Still later in the century when the activities of the French in Mexico violated the Monroe Doctrine, the threat was met not by cutting the maritime communications between France and Mexico, but rather by massing Sherman's veterans along the Rio Grande. American power was thus virtually never utilized outside the American continents during this period and was confined to the gradual elimination of all potential threats to American security which might originate within that Hemisphere. This phase may be said to have come to an end with the final pacification of the Indians in the 1890s and its termination is symbolized in Olney's bold statement to the

British government during the 1895 Venezuela boundary dispute, "Today the United States is practically sovereign on this continent, and its fiat is law upon the subjects to which it confines its interposition."

The Navy's subordinate role during this Continental Phase of policy is well indicated by the miscellaneous nature of its military functions. These were basically threefold. First, there were the Navy's responsibilities for coastal defense. From the time of Jefferson's administration down through the 1880s this resulted in the construction of a whole series of gunboats and monitors designed solely for this purpose. Secondly, the Navy was responsible for protecting American commerce overseas and, in the event of war, raiding the commerce of the enemy. For this purpose the Navy was deployed in half a dozen squadrons scattered about the world from the Mediterranean to the East Indies and was largely equipped with fast frigate cruiser type vessels. Thirdly, during the Mexican War and the Civil War, when the United States was fighting two nations powerless at sea, the Navy performed valuable functions in blockading the enemy and assisting in amphibious operations. These miscellaneous military functions did not, however, exhaust the activities of the Navy during this period. Since these military functions were of a general secondary nature, the Navy tended to acquire a wide variety of essentially civilian functions not directly related to any security threat. These included the support of general scientific research, the organization of a number of exploring expeditions, the frequent performance by naval officers of diplomatic functions, and the utilization of members of the naval service to administer civilian departments of government. In general, during this period the Navy had no clearly essential role to play in meeting any major security threats and consequently tended to dissipate its energies over this wide variety of civilian and military functions.

The subordinate role of the Navy in implementing national policy was reflected in the weak public support which it received during this period. The continuous expansion of the nation westward tended steadily to decrease the political power of those sections most sympathetic to the Navy, and after the Federalists were swept out of office in 1800 it is not inaccurate to say that the government was generally dominated by political groups either indifferent to or actively hostile towards the Navy. The farmers of the interior tended to view the naval establishment as an unnecessary if not dangerous burden on the national economy. Consequently the Navy was frequently allowed to fall into fairly serious states of disrepair, reaching its lowest point in the post Civil War years.

Since the Navy had no definite role to play in implementing national policy, it was unnecessary for it to have a type of organization which emphasized a distinction between its military and civilian functions. Consequently, although there was a major change in naval organization in 1842, when the bureau system was introduced, nonetheless the basic pattern of naval organization remained the same throughout this entire period. Neither under the Board of Naval Commissioners nor under the bureaus was there any clear differentiation between the military and the civilian functions of the naval department under the supervision of the Secretary. When during the Civil War the Navy was called upon to perform a significant military function, a special officer had to be designated to direct the military activities of the fleet. With this exception, however, naval organization reflected the inability of the Navy to develop a strategic concept relating it to the goals of national security policy.

IV

The Navy and National Policy: Oceanic Phase

All this changed in the 1890s when the United States began to project its interests and power across the oceans. The acquisition of overseas territorial possessions and the involvement of the United States in the maintenance of the balance of power in Europe and Asia necessarily changed the nature of the security threats with which it was concerned. The threats to the United States during this period arose not from this continent but rather from the Atlantic and Pacific oceanic areas and the nations bordering on those oceans. Hence it became essential for the security of the United States that it achieve supremacy on those oceans just as previously it had been necessary for it to achieve supremacy within the American continent. This change in our security policy was dramatically illustrated by the war with Spain. What began as an effort to dislodge a secondary European power from its precarious foothold on the American continent ended with the extension of American interests and responsibilities to the far side of the Pacific Ocean.

This new position of the United States made it one of several major powers each of which was attempting to protect its security through the development of naval forces. This meant dramatic changes in the position of the Navy, and the role of the Army in implementing national policy became secondary to that of the Navy. Instead of performing an assortment of miscellaneous duties none of them particularly crucial to the national security, the Navy was now the Nation's "First line of defense." In a little over twenty years, from 1886 down to 1907, the United States Navy moved from twelfth place to second place among the navies of the world. This dramatic change required a revolution in the thinking of the Navy, the operations of the Navy, and the composition of the Navy.

The revolution in naval thinking and the development of a new strategic concept for the Navy reached its climax, of course, in the work of Alfred Thayer Mahan. The writings of this naval officer accurately portrayed the new role of the Navy. Attacking the old idea that the functions of the Navy were related to coastal defense and commerce destruction, Mahan argued that the true mission of a navy was acquiring command of the sea through the destruction of the enemy fleet. Mahan vented his scorn upon the "police" functions to which the Navy had been relegated during this previous period of national strategy. Writing at a time when national strategy was undergoing a profound change, he failed to realize that these "police" functions had been just as well adapted to the achievement of national aims in this period as his "command of the sea" doctrine was in the new age which was just beginning. To secure command of the sea it was necessary to have a stronger battlefleet than the enemy. This could only be secured by building more ships than other nations, insuring that the ships which one did build were larger and had more fire power than those of other nations, and keeping those ships grouped together in a single fleet instead of deployed all over the world in separate squadrons. The net results were naval races, big-gun battleships, and the theory of concentration as the chief aim of naval strategy.

As generalized in the preceding paragraph, the Mahan doctrine was accepted by virtually all the world's naval powers. Each country, however, also had to apply the doctrine to the threats peculiar to it. Down until World War I the United States was about equally concerned with the threats presented by the Japanese and German navies. The fleet was kept concentrated on the Atlantic coast—this was the location of most of the shipyards and the Navy's most consistent public support—and the Isthmus canal was rushed to completion. With the destruction of German surface power the fleet was shifted to the Pacific, and throughout the following two decades American naval thought was oriented almost exclusively towards the possibility of a war with Japan. This was responsible not only for the location of the fleet but also for the development of weapons and techniques which could be effectively employed in the broad reaches of the Pacific. In the 1941–1945 naval war with Japan, the Navy in effect realized the strategic concept which dominated its planning for twenty years.

The increased importance of the Navy to national security towards the end of the nineteenth century was paralleled by the increased prestige of the Navy throughout the country. Public opinion came to view the Navy as the symbol of America's new role in world affairs. Business groups which were now playing an increasingly important role in government were generally more favorably inclined towards the Navy than the agrarian groups which had previously been dominant. The Navy League of the United States was organized and played a major role in interpreting the Navy to the public. Presidents—particularly the two Roosevelts—and congressional leaders turned a more sympathetic ear to the Navy's requests for funds. Thus the Navy was able to get that public support which was necessary for it to implement its strategic concept.

The emergence of a well-defined military function for the Navy meant that the old organization of the Navy Department had to be altered also. The formation of the fleet and the development of its purely military role permitted the business of the Department to be roughly divided into the two categories of military functions and civilian functions. The reformers within the Navy hence campaigned for an organizational structure which reflected this duality of function. This campaign resulted in the creation of the General Board in 1900, the institution of the naval aids in 1909, and eventually the creation of the Office of Naval Operations in 1915. In time, the Chief of this latter office assumed the responsibility for the military aspects of the Navy while the bureau chiefs continued to report directly to the Secretary on the performance of their civilian duties.

V
National Policy in the Eurasian Phase

The close of World War II marked a change in the nature of American security policy comparable to that which occurred in the 1890s. The threats which originated around the borders of the Atlantic and Pacific Oceans had been eliminated. But they had only disappeared to be replaced by a more serious threat originating in the heart of the Eurasian continent. Hence American policy moved into a third stage which involved the projection, or the possible projection in the event of war, of American power into that continental heartland. The most obvious and easiest way by which this could be achieved was by long-range strategic bombing and consequently American military policy in the immediate postwar period tended to center on the atomic bomb and the intercontinental bomber. Subsequently the emphasis shifted to the development of a system of alliances and the continuing application of American power through the maintenance of United States forces on that continent. These two approaches furnished the Air Force and the Army with strategic roles to play in the implementation of national policy. What, however, was to be the mission of the Navy? How could the Navy play a role in applying American power to the Eurasian continent? This was the challenge which the new dimension of American foreign policy placed before the Navy, which temporarily caused the Navy to falter and hesitate, and which finally was met by the development of a New Naval Doctrine defining the role of the Navy in the Cold War.

VI
The New Naval Doctrine: The Transoceanic Navy

This new doctrine as it emerges from the writings of postwar naval writers and leaders basically involves what may be termed the theory of the transoceanic navy, that is, a navy oriented away from the oceans and towards the land masses on their far side. The basic elements of this new doctrine and the differences between it and the naval concept of the Oceanic phase may be summarized under the headings that follow.

1. THE DISTRIBUTION OF INTERNATIONAL POWER

The basis of the new doctrine is recognition of the obvious fact that international power is now distributed not among a number of basically naval powers but rather between one nation and its allies which dominate the land masses of the globe and another nation and its allies which monopolize the world's oceans. This bipolarity of power around a land-sea dichotomy is the fundamental fact which makes the Mahanite concept inapplicable today. For the implicit and generally unwritten assumption as to the existence of a multi-sea power world was the foundation stone for Mahan's strategic doctrine. Like any writer Mahan grasped for the eternal verities and attempted to formulate what seemed to him the permanent elements of naval strategy. But also like every other writer his theory and outlook were conditioned by the age in which he lived. That age was one in which the decisive wars were between competing naval powers. This multi-sea power world had its origins in the rise of the European nation-state system, the discovery of the New World, and the resulting competition between the European nations for overseas colonies and trade. This period of sea power competition lasted roughly from the middle of the seventeenth century to the middle of the twentieth and is divisible into two sub-periods. The first sub-period lasting to 1815 was characterized by intense naval competition and warfare between Spain, the Netherlands, France, and Great Britain. In the end, after the series of exhausting conflicts culminating in the Napoleonic Wars and Trafalgar, Great Britain emerged as the dominant sea power. From 1815 down to the 1890s she maintained this position without serious challenge. By the end of the century, however, a new round of competition developed as Germany, the United States, and Japan arose to challenge British naval supremacy. This second period witnessed the defeat of the German and Japanese navies in World War I and World War II respectively, and ended with Anglo-American, or, more specifically, American naval power dominant throughout the world.

In the light of this naval history it is important to recognize that Mahan's entire thought was geared to this sea power stage in world history. Basically what he did was to study intensively the first sub-period in this stage and then apply the principles gained from such study to the second sub-period in which he lived. This technique gave a superficial air of lasting permanence to his doctrine: for if the principles underlying seventeenth century naval warfare and sea power were applicable at the end of the nineteenth century, then surely these must be universal principles valid throughout history. In actuality, these two sub-periods were, however, unique in their similarity. The first coincided with the initial surge of European colonialism into the New World, and the second coincided with the later surge of that colonialism into Africa and Asia. These are not situations which will be repeated again.

It should also be noted that it was not just chance which led Mahan to concentrate his historical studies on the period from 1660 to 1815. For, although he admitted in a letter to Rear Admiral Stephen A. Luce that "there are a good many phases of naval history," he nonetheless believed that he had been "happily led to take up that period succeeding the peace of Westphalia, 1648, when the nations of Europe began clearly to enter on and occupy their modern positions, struggling for existence and predominance." And it was also generally characteristic of this period that, as Mahan said, except for Russia and possibly Austria, the force of every European state could "be exerted only through a navy."

All the other facets of Mahan's thought rest upon his assumption of the existence of two or more competing naval powers. The idea that the purpose of a navy is to secure command of the sea, that to achieve this end concentration of force in a battlefleet is necessary, and that victory will go to that fleet with the biggest ships, the biggest guns, and the thickest armor, all rest logically on this premise. For obviously the concentration of force in a battlefleet is necessary only if the enemy is capable of doing the same. And, as Bernard Brodie has pointed out, the idea of developing a battlefleet to secure command of the sea originated in the Anglo-Dutch Wars of the middle seventeenth century, at the beginning of this sea power phase of history.

To deny the permanent validity of Mahan's theory is not to deny the brilliance of Mahan's insight. To describe and formulate the principles underlying the major developments in world history over a period of three hundred years is no mean achievement. But we must not permit the impressiveness of Mahan's accomplishment to blind us to the inapplicability of his strategic concept at the present time. A world divided into one major land power and one major sea power is different from a world divided among a number of rival sea powers. The strategy of monopolistic sea power is different from that of competitive sea power. The great oceans are no longer the no man's land between the competing powers. The locale of the struggle has shifted elsewhere, to the narrow lands and the narrow seas which lie between those great oceans on the one hand and the equally immense spaces of the Eurasian heartland on the other. This leads us to the second element which distinguishes the new strategic doctrine from the old.

2. THE SITE OF DECISIVE ACTION

The Mahan theory justly emphasized not only the influence of sea power but also the decisiveness of naval battle. The sea was a battleground, "a wide common," and the only avenue through which every power could strike at the interests of every other power. Major fleet actions were the decisive events in most of the principal wars of this period from the defeat of the Spanish Armada in 1588 to the dispersion of the remnants of the Japanese Fleet in the Battle of the Philippine Sea in 1944. Between these encounters there were a whole series of naval battles which significantly influenced the course of history: Lowestoft, The Texel, Beachy Head, Ushant, Trafalgar, Manila Bay and Santiago, Tsushima Straits, Jutland, Coral Sea, Midway. Mahan demonstrated the decisive character of the naval engagements in the first round of naval competition; and his teachings and his successors have illuminated the decisiveness of the subsequent ones. While not denying the importance of land battles, nor the significance of such techniques as naval blockade, the strategic concept of this previous age nonetheless emphasized the significance of naval engagements fought solely at sea.

In a world in which a continental power confronts a maritime power, this is no longer possible. As most recent naval writers have recognized, major fleet actions are a thing of the past. The locale of decisive action has switched from the sea to the land: not the inner heart of the land mass, to be sure, but rather to the coastal area, to what various writers have described variously as the Rimland, the Periphery, or the Littoral. It is here rather than on the high seas that the decisive battles of the Cold War and of any future hot war will be fought. Consequently, naval writers in the period since 1945 have not hesitated to admit and, indeed, to proclaim the importance of ground force. The reduction of enemy targets on land, Admiral Nimitz stated, "is the basic objective of warfare." Criticizing the Mahan doctrine for tending to erect sea power into an independent thing-in-itself (a view which was not far wrong when the conflict of sea power against sea power was the decisive event in war), Walter Millis argues that:

> Korea is one long lesson in the double fact that all military power is "land power"; and that it can be effectively exercised, under the conditions created by modern technology, only by the most skillful combination and concentration of all available weapons, whether airborne, seaborne, or earthborne to achieve the desired political ends under the particular circumstances which may arise.

3. THE MISSION OF THE NAVY

This fact that decisive actions will now take place on land means a drastic change in the mission of the Navy. During the previous period, this mission was to secure command of the sea. "[I]n war," Mahan said, "the proper objective of the navy is the enemy's navy," and as he further remarked in another classic passage:

> It is not the taking of individual ships or convoys, be they few or many, that strikes down the money power of a nation; it is the possession of that overbearing power on the sea which drives the enemy's flag from it, or allows it to appear only as a fugitive, and which, by controlling the

great common, closes the highways by which commerce moves to and from the enemy's shores. This overbearing power can only be exercised by great navies....

Since the American navy now possesses command of the sea, however, and since the Soviet surface navy is in no position to challenge this except in struggles for local supremacy in the Baltic and Black Seas, the Navy can no longer accept this Mahanite definition of its mission. Its purpose now is not to acquire command of the sea but rather to utilize its command of the sea to achieve supremacy on the land. More specifically, it is to apply naval power to that decisive strip of littoral encircling the Eurasian continent. This means a real revolution in naval thought and operations. For decades the eyes of the Navy have been turned outward to the ocean and the blue water; now the Navy must reverse itself and look inland where its new objectives lie. This has, however, been the historical outlook of navies which have secured uncontested control of the seas, and as Admiral Nimitz has pointed out, during the period of British domination, "it is safe to say that the Royal Navy fought as many engagements against shore objectives as it did on the high seas." It is a sign of the vigor and flexibility of the Navy that this difficult change in orientation has been generally recognized and accepted by naval writers and the leaders of the naval profession.

The application of naval power against the land requires of course an entirely different sort of Navy from that which existed during the struggles for sea supremacy. The basic weapons of the new Navy are those which make it possible to project naval power far inland. These appear to take primarily three forms:

(1) carrier-based naval air power, which will in the near future be capable of striking a thousand miles inland with atomic weapons;
(2) fleet-based amphibious power, which can attack and seize shore targets, and which may, with the development of carrier-based air lifts, make it possible to land ground combat troops far inland; and
(3) naval artillery, which with the development of guided missiles will be able to bombard land objectives far removed from the coast.

The navy of the future will have to be organized around these basic weapons, and it is not utopian to envision naval task forces with the primary mission of attacking, or seizing, objectives far inland through the application of these techniques.

4. THE BASE OF THE NAVY

In the old theory the sea was the scene of Operations and navies consequently had to be based on land. In the ultimate sense that is still true since man must still draw his sustenance and materials from land. But it is also possible to argue that the base of the Navy has been extended far beyond the limits of the continental United States and its overseas territorial bases. For in a very real sense the sea is now the base from which the Navy operates in carrying out its offensive activities against the land. Carrier aviation is sea based aviation; the Fleet Marine Force is a sea based ground force; the guns and guided missiles of the fleet are sea based artillery. With its command of the sea it is now possible for the United States Navy to develop the base-characteristics of the world's oceans to a much greater degree than it has in the past, and to extend significantly the "floating base" system which it originated in World War II. The objective should be to perform as far as practical the functions now performed on land at sea bases closer to the scene of operations. The base of the United States Navy should be conceived of as including all those land areas under our control and the seas of the world right up to within a few miles of the enemy's shores. This gives American power a flexibility and a breadth impossible of achievement by land-locked powers.

The most obvious utilization of this concept involves its application to carrier aviation. In the words of Admiral Nimitz:

> The net result is that naval forces are able, without resorting to diplomatic channels, to establish off-shore, anywhere in the world, airfields completely equipped with machine shops, ammunition dumps, tank farms, warehouses, together with quarters and all types of accommodations for personnel. Such task forces are virtually as complete as any

air base ever established. They constitute the only air bases that can be made available near enemy territory without assault and conquest, and furthermore, they are mobile offensive bases that can be employed with the unique attribute of secrecy and surprise, which contributes equally to their defensive as well as offensive effectiveness.

From this viewpoint it is possible to define the relation of the Navy's important anti-submarine responsibilities to these newer functions. Submarine warfare is fundamentally a raiding operation directed at the Navy's base. If not effectively countered, it can of course have serious results. But A.S.W., although vitally important, can never become the primary mission of the Navy. For it is a defensive operation designed to protect the Navy's base, *i.e.*, its control and utilization of the sea, and this base is maintained so that the Navy can perform its important offensive operations against shore targets. Antisubmarine warfare has the same relation to the Navy as guarding of depots has for the Army or the protection of its airfields and plane factories has for the Air Force. It is a secondary mission, the effective performance of which, however, is essential to the performance of its primary mission. And, indeed, the successful accomplishment of the primary mission of the Navy—the maintenance of American power along the littoral—will in itself be the most important factor in protecting the Navy's base. For holding the littoral will drastically limit the avenues of access of Soviet submarines to the high seas.

5. THE GEOGRAPHICAL FOCUS OF NAVAL OPERATIONS

This new theory of the transoceanic navy differs from the old Mahanite doctrine in that its principles are applicable to only one Navy instead of several. We have seen how each nation had to adopt the old Mahanite theory to its own specific circumstances, and for the United States this eventually meant focusing its attention upon the Pacific Ocean. Is there any such specific geographical area which assumes special importance in the application of the new theory? Obviously this theory applies in general to the entire littoral of the Eurasian continent from Kamchatka to the North Cape (and especially to peninsulas such as Korea).

Even a superficial glance at the map of Eurasia, however, will reveal that there is one area which specially lends itself to offensive naval operations against the land. This is, of course, the Mediterranean Basin. For, in effect, the Mediterranean extends the base of American power 2,500 miles inland into the Eurasian continent. From this basin naval power can be projected over most of Western Europe, the Balkan peninsula, Turkey, and the Middle East. In the event of a major war with Russia, the Mediterranean would be the base from which the knock-out punch could be launched into the heart of Russia: the industrial-agricultural Ukraine and the Caucasus oil fields. It is consequently hardly surprising to find that the Mediterranean has now replaced the Pacific as the geographical focus of attention for the American Navy.

The recognition of the crucial role of the Mediterranean Basin in the implementation of American foreign policy can be dated from the historic announcement by Secretary Forrestal on September 30, 1946, that American naval forces would be maintained in that area for the support of our national policy. The increase in the strength of these forces and the creation of the Sixth Task Fleet on June 1, 1948, were further steps in the implementation of this policy. The carrier aviation, surface power, and amphibious forces of this fleet have been recognized as being of crucial importance in supporting American policy in this area. This key role of the Mediterranean has been reflected in the attention devoted to it in naval writings, and it has even been described as the "sea of destiny"—a term previously reserved for the Pacific Ocean. This concentration of attention upon the Mediterranean does not, of course, mean that the application of naval power will not be important at other points along the littoral. But it does mean that at least for the foreseeable future the Mediterranean offers the most fruitful area for the Navy's performance of its new function.

6. THE AIM OF NAVAL TACTICS

Under the old theory it was necessary to concentrate naval forces in order to win control of the sea. Consequently the battlefleet emerged as the main instrument of sea power. Now, however, concentration is necessary at or over the target on land, and hence for defensive purposes dispersion and deception are essential for the fleet at sea. Planes from a number of widely separated carriers can, for instance, be concentrated over their target and secure local air supremacy there. Only in amphibious landings would any large-scale concentration of naval vessels be necessary and even there new techniques may avoid the massing of a large number of ships in a small area. Since these new functions permit the Navy to avoid concentrating its ships afloat, there is consequently little basis for the argument that the effectiveness of atomic bombs against a concentrated fleet has ended the usefulness of the Navy. Dispersion, flexibility, and mobility—not concentration—are the basic tactical doctrines of the new Navy.

VII
Public Support and Naval Organization

Inevitably a new strategic concept must have significant implications for the Navy's public support and its organizational structure. So far as the latter is concerned the implications of this concept are as yet difficult to identify. Certainly once there is general acceptance of the new role of the Navy, the Navy will be able to afford to take a more favorable attitude to further unification of the armed services. Certainly also a recognition of this new function should eventually find its way into law since the National Security Act still defines the primary mission of the Navy as "prompt and sustained combat incident to operations at sea." In general, it is probable that the dual basis of naval organization developed during the Oceanic phase can continue to be the basis of naval organization. In any case, it is likely that the most important implications of the new doctrine involve public support rather than organization.

Perhaps the first necessity of the Navy with respect to this is for it to recognize that it is no longer the premier service but is one of three equal services all of which are essential to the implementation of American Cold War policy. The second necessity is for the Navy to insist, however, upon this equal role. To maintain its position the Navy must develop public understanding of its transoceanic mission. As it is now, the experts on military affairs—columnists such as Hanson Baldwin and Walter Millis—thoroughly appreciate the Navy's role, but too often one still hears from the average American the question: "What do we need a navy for? The Russians don't have one." This attitude can only be overcome by a systematic, detailed elaboration and presentation of the theory of the transoceanic Navy against the broad background of naval history and naval technology. Only when this is done will the Navy have the public confidence commensurate with its important role in national defense.

4 "MAKE THE WORD BECOME THE VISION"

CAPT Peter M. Swartz, USN, and CAPT John L. Byron, USN

The **November 1992** issue of *Proceedings* included a Navy–Marine Corps "White Paper" entitled " . . . From the Sea: Preparing the Naval Service for the 21st Century" that offered a "new direction of the Navy and Marine Corps team" described as "Naval Expeditionary Forces—Shaped for Joint Operations—Operating Forward from the Sea—Tailored for National Needs." In that same issue, Captain Swartz and Captain Byron offered a companion piece that "laid the groundwork for the expansion and elaboration of the White Paper's central themes."

"MAKE THE WORD BECOME THE VISION"

By CAPT Peter M. Swartz, USN, and CAPT John L. Byron, USN, U.S. Naval Institute *Proceedings* (November 1992): 71–73.

How does the Navy get its message across? Here's a simple two-part answer:

- Have a message.
- Get it across.

But what message? How to craft it? And how to get it across? That's not so simple—but we have some solid clues about where to start. Let's begin with two questions:

- What has worked before?
- What is working now?

What has worked before is a single, cohesive, carefully constructed widely disseminated strategic vision of the Navy, used both to inform and to drive the Navy's future—both inside and outside the organization. Our last successful presentation of such a central vision was the Maritime Strategy of the 1980s. Earlier themes from Admirals Forrest Sherman and Elmo Zumwalt had attempted the same approach, but now the stage looks bare, the vision is missing, and many are freely speaking their minds. Today's cacophony must give way to a single voice, a single strong argument, and a formal presentation—to serve as the touchstone of Navy strategy, planning, resourcing, marketing, design, and operations. Without a doubt, we need to emulate the extraordinary success of the Maritime Strategy.

What is working now? Those "single, cohesive, carefully constructed and widely disseminated strategic visions" currently extant: National Military Strategy and Base Force of the Secretary of Defense and Chairman of the Joint Chiefs of Staff; the Air Force's Global Reach, Global Power, and the Army's update of Air Land Battle—all of these work. At present we have no Navy equivalent. We have a marvelous Navy, absolutely vital to the existence of the nation, but we have not been able to say to ourselves—nor to anyone else—what the U.S. Navy does, what it should look like, or why we even need a navy. In the ongoing contest for roles and missions we are getting our butts kicked.

We need a message. But not just any message will do, nor is its crafting merely a minor detail. Our message should have certain characteristics, and it should be built in a way that ensures its survival:

- *Give the message a name.* Keep the same name all the time. Let the message evolve and the messengers change, but keep the name. No name will be perfect, but changing buzz words every few months is ludicrous.

- *Leadership is the critical component.* A marvelous descriptive phrase for what we need is "rhetorical leadership." Our message must come from the top down, and it must be repeated again and again by both our uniformed and civilian leaders, singing in chorus. Any question from any source must get the same clear answer. Every time a Navy leader gets in front of a microphone or picks up a pen, the same message has to come out. No message—even a brilliant one—can endure without the full commitment of all the Navy's leadership.
- *Build the message the right way.* This thing can't be put together by a committee or a bureaucracy. Every time we try this, we get pablum. The way we've built sound messages and plans, from the beginnings of War Plan Orange to the full flower of the Maritime Strategy, has first been to task a few smart, experienced mid-grade officers: to draft the product; to pulse simultaneously the fleets, schools, and staffs; and then personally to move it up a very short, vertical chain all the way to the top. Then the Navy's top leaders must sell it collectively and gather support for it among senior leadership of those same fleets, schools, and staffs, including those of the Marines.
- *Package the message properly.* We need a secret version, to show the insiders we're serious. We also need an unclassified version, to shape the debate everywhere we can within the Navy and outside the service. (Admiral Sherman's failure lay in not presenting his message outside of a tight security circle. Our message should be the engine pulling the Navy forward. This won't happen in a vacuum.)
- *Keep the message simple.* We have a lot to put in this message but it has to be understandable on its own terms. This argues for simplicity, as we had in the basic outline of the Maritime Strategy. Eschew obfuscating prolixity and gratuitous redundancy—please!
- *Don't get bogged down in hardware.* The beauty of the Maritime Strategy was its use of "warfare areas" as the basic coin, with platform-think subordinated. Geography, geopolitics, global interests, national character, the military value of freedom of the sea, and the unique nature of war at sea and from the sea are proper elements of the message, either implicit or stated explicitly. Keep the hardware on the sidelines.

- *Be resonant with shared naval concepts.* The March 1992 *Proceedings* article "It's War With Anastasia" attempted to capture in a structured way the shared beliefs that Navy officers bring to strategy and warfare. Such themes are broadly understood in the Navy, and should form the foundation on which the message is built.
- *Be resonant with history.* This period of Navy confusion is not unprecedented. The historical record suggests that the clear-threat, clear-response character of the Navy in the Cold War was much rarer than today's milling about. But the Maritime Strategy, as a vehicle for the Navy message, itself had precedents. Mahan's polemics created the modern Navy and gave us shared values that persist today. War Plan Orange corralled Navy thought for at least 35 years. The visions of Admiral Sherman and his contemporaries set the Navy on course for the Cold War. In this century, the Navy always has had a message—or has had the need to articulate one. We need to articulate one right now. History can help us start.
- *Speak with one voice.* In crafting our message and getting it out, we will be forced to deal with the most pernicious evil of today's Navy: communityism. When it was raining nickels and there was plenty of money for everyone, platform primacy did not affect survival. But now in this time of scarcity we face a certain choice: either community or Navy. The answer for the United States must be "Navy," of course, but that need not submerge totally our community interests or identity. Instead we need to do within the Navy what Goldwater-Nichols has done among the services. To use a wonderful term from a Naval Postgraduate School lieutenant, we need *inner jointness*, which advocates a cross-functional approach (recognize the term from Deming?) that subordinates community themes to a broader bluesuit understanding. If we continue to put community interests above Navy, we are dead. The message that each community needs is the one we should be crafting; a single Navy message for one Navy.
- *Answer the critics; address the automatic doubts.* Critics will have two strong comments. One is that the Navy is too fractionalized for anyone to believe a putatively cohesive single message from the Navy. The answer to that is

inner jointness, which puts all the communities and all the leaders in harness. The second certain criticism is that—whatever our words—the Navy really intends to go it alone. To respond to this we need a distinctly joint message. Jointness is here to stay. America wants its Navy to fight jointly, so every paragraph must tip its hat to jointness. We have to demonstrate the truth: that we are part of the overall defense, that we support the broad defense interests of the nation, that we are fully integrated with and part of the overall military strategy; and that we complement the other services, not rival them.

- *Be resonant with the current winning message in defense.* The theme of the Maritime Strategy was defeat of the Soviets at all levels of conflict. It meshed perfectly with overall defense policy, intelligence estimates, and national strategy. Its themes were absolutely resonant with the messages of higher echelons and those of sister services and agencies. We always gain strength when we can link to higher messages, to prove what we're saying is worth hearing in broad context. Seamless coherence with overall defense themes also is a necessary characteristic of the message. (The articles, testimony, and speeches of Secretary of Defense Dick Cheney and Chairman Colin Powell of the Joint Chiefs of Staff offer a strong architecture. General Powell's vision is an especially lucid argument for a sound defense plan. We should seriously consider adopting the structure and vocabulary of these formulations in developing the Navy message. Dr. Jim Tritten showed us how in July's *Submarine Review*.)

- *Anticipate potential changes to the current winning message in defense.* But let's not link it too closely to higher themes. We'll need to detune the Navy message ever so slightly, so it doesn't depend critically on politics or the precise formulation it's tied to, and so we can let our message evolve to stay with changing world conditions—including changes in the senior defense leadership. We must avoid being torpedoed by our own message, as we would be if we failed to build in a certain amount of flexibility and resilience.

- *Stay the course.* Our message should reflect broad and enduring naval themes, not the narrow views of a specific set of incumbent Navy leaders (recall that the Maritime Strategy advanced through at least three Secretaries of the Navy and as many Chiefs of Naval Operations). The Army's successful and continuously updated message, Air-Land Battle, has been around since 1982. It is institutionalized in the field manual *FM 100-5* and in every training and education course in the Army. We must do the same with our message. We need to write a "keeper." Then we need to keep it until world events and national security policy mandate a change.

- *Get everybody on board.* Internal marketing and tight discipline are paramount to success; we must bring the entire Navy leadership into support of this message. In practical terms, this means that we let all leaders provide inputs and chop on it—coordinating with, convincing, and coopting the fleet commanders, the Marine Corps leadership, the Unified commanders, and the new OpNav staff before we go final.
- *Market the message.* We need to push the message into every place that can support it. We need: to get the Navy informed about the game plan and the need for all to get on board; to connect with key congressional leaders, Hill staffers, media elites, academics, defense writers, et al., to focus external thinking on the message; to institutionalize the message in Navy schools, service colleges, training courses, warfare publications, and war plans; and to be singleminded in pushing this vision in joint dealings at both CinC and JCS levels, to wire the message into higher-level plans and documents.
- *Drive everything with the message.* Is the Navy message a strategy, a force-planning guide, a resource allocation model, a research and development template, an Operational scheme, a public-relations message, or a congressional argument? Yes. Our message, beginning with strategy, becomes the way all other arguments are made. It's a tiebreaker. It's the theme and the vision for all our planning. It's the way we decide where money goes and what the money goes for—so it's the way the smart guys argue for their programs. It's the reinforcing wraparound for all our dialogue with the other services, the Joint Staff, Congress, Office of Secretary of Defense, and the public.
- *Protect the message.* One beauty of the Maritime Strategy was that the entire Navy felt the ownership of it. We need that spirit for the new message, too—

a protective sense that this is our collective statement and that we should not undercut it by shooting at our own feet. But what of the improvements, the evolutionary refinements to the message that we experienced with the Maritime Strategy? Yes, we will need constant review and forward movement because this should be a living message, and continually adding better ideas is important. But we must move it forward very carefully—not at the expense of Navy solidarity. Free-lancing will kill us. The way to improve the message is to improve it internally, and not have senior officers argue with it in public or challenge it with a different product. The various elements of the Navy should let the message evolve through a dynamic but orderly process of continuous improvement, resident in the OpNav staff. We must resolve our problems with the message from within the system and maintain a united front to the outside world. Here again, leadership will be critical.

- *Only one message.* It seems that the entire naval establishment is in the message business today—with nearly every Navy power center and bicycle shop in the Pentagon and the Fleet writing its own version of what the Navy should be saying. Fleet commanders float their trial balloons. Senior speechwriters labor independently. The submarine force has produced a new vision statement of its own. The Total Quality Leadership office has a vision statement. The CNO Executive Panel has published a Navy Policy Book with little fanfare. Deputy Chief of Naval Operations for Plans, Policy and Operations (OP-06) has toiled for years to get out its version of the message, with drafts cascading upon drafts. Other baronies have circulated their own rival claimants. Let's face it: we are building a Tower of Babel—but there's no architecture. Little thought connects one document to the next. Each seems to have been written on its own planet. The authors come from tribes that have seldom met, except in battle.

- *Timing is everything.* Navy Secretary John Lehman launched public discussion of the Maritime Strategy at the very start of the Reagan-Weinberger-Lehman era. General Powell had the Base Force Concept on the street by the end of his first year in office, *Joint Pub 1* on the street at the end of his second year, and the National Military Strategy out with almost two years left on his watch. Both had lots of time to repeat—and institutionalize—their messages. Our timing was fine on "The Way Ahead" article (*Proceedings*, April 1991), but the *Navy Policy Book* had the bad luck to be signed out by Secretary of Navy H. Lawrence Garrett just months before he left office. It can do the Navy little good to launch its message in the midst of a major convulsion—such as Tailhook—or shortly before major changes in our National Command Authority or Navy leadership. But if that's when the message must be released, then the requirements to be loud, consistent, and repetitive become all the more powerful. And any document with yesterday's signatures on it will have to be updated, turned around, and put back out on the street immediately, if it is to have lasting effect.

To sum up—multiple messages are no message at all. We must form ranks and march as one Navy, with one message and one vision. Our uncertain future hangs on it.

5 "CRAFTING A 'GOOD' STRATEGY"

LtCol Frank Hoffman, USMCR (Ret.)

The quest for a modern maritime strategy has been an ongoing challenge for decades. Here, Colonel Hoffman warns that the term "strategy" may have lost its true meaning as we take that quest into the twenty-first century. To help refocus that quest, he provides a pragmatic approach that includes the elements of good strategy (A Diagnosis, A Guiding Policy, and Coherent Action) and nine "Common Strategy 'Sins'" that include ignoring a scarcity of resources, mistaking strategic goals for strategy, and not recognizing that implementation is part of the process. He concludes by warning that a "good" strategy is "not a vision or ephemeral statement of the obvious" but instead "involves hard choices, clear objectives, a continuous assessment of risks, and priorities."

"CRAFTING A 'GOOD' STRATEGY"

By LtCol Frank Hoffman, USMCR (Ret.), U.S. Naval Institute *Proceedings* (April 2012): 58–63.

There's no time like the present to start looking into the proverbial crystal ball and thinking about a maritime strategy "refresh."

Has the term "strategy" completely lost its meaning in the U.S. defense community? Some scholars bemoan the overall decline or lack of capacity for strategic thinking in the United States.[1] The situation appears compounded by a weak grasp of what a true strategy is, and a thin understanding of the process by which strategy is formulated and implemented. More specific to this audience, Navy Commander Michael Junge asserts that the service no longer has a culture of strategic thought.[2]

Chief of Naval Operations Admiral Jonathan Greenert has tasked the naval strategic community to assist him in updating the *Cooperative Strategy for 21st Century Seapower* (hereafter referred to as *CS-21*). Commander Junge characterizes the recommendation to reassess *CS-21*, only four years after its release, as further proof of a strategy problem. He argues that we lack "a consistent, long-term, cohesive, and followed strategy." Taking the opposite tack, this is actually an excellent time to adapt our strategy and a perfect opportunity to clarify what a maritime strategy should contain and how it should be applied. A good strategy, like proper operational planning, must adapt to the environment.

Strategic Adaptation

The noted U.S. military historian Dr. Williamson Murray, now teaching at the Naval War College, once directed a major study on the making of strategy. This monumental effort, examining some 17 cases spanning several centuries, produced the definitive foundation for understanding how states and bureaucracies go about framing their strategic intentions and applying them in the real world. Instead of a static, consistent, long-term application of a single document, Murray concluded that strategy is best viewed as a process. He and his fellow historians found that the process does not conclude with a fixed product but instead "a constant adaptation to shifting conditions and circumstances in a world where chance, uncertainty, and ambiguity dominate."[3]

This conclusion matches with experience in the corporate world. Henry Mintzberg, an international management expert, observed that the whole nature of strategy-making was "dynamic, irregular, discontinuous, calling for groping, interactive processes with an emphasis on learning and synthesis."[4] He found the real world too complex to allow strategies to take shape by planners in the form of clear plans or fixed visions. He noted that many organizations develop deliberate plans on paper, but what really happens is that learning occurs as these plans meet with the dynamics of a changed operating environment, new competitors, unexpected technological breakthroughs, etc. Thus, he argued that strategies must emerge in small steps, as an organization adapts, or "learns."[5]

Good vs. Bad Strategy

We should make a differentiation between good strategies and bad. Bad strategies are far more common, as defined by UCLA professor Richard Rumelt, who notes:

> Having conflicting goals, dedicating resources to unconnected targets and accommodating incompatible interests are the luxuries of the rich and powerful, but they make for bad strategy. Despite this, most organizations will not create focused strategies. Instead, they will generate laundry lists of desirable outcomes, and ignore the need for genuine competence in coordinating and focusing their resources.[6]

To Rumelt, good strategy is "coherent action backed by an argument. And the core of the strategist's work is always the same: discover the crucial factors in a situation and design a way to coordinate and focus actions to deal with them."[7] This must be the touchstone for the next iteration of our maritime strategy.

Strategy is about choices, objectives, priorities, risk, and coherence—which is a logical thread between ends, ways, and means. American strategy repeats a number of faults all too frequently. The intellectual foundation for policy aims or strategic plans are often thin or inadequately challenged. Sometimes U.S. strategy is simply a list of unprioritized objectives or alliterative bullets on a PowerPoint slide. Sometimes our strategies are mere statements of a strategic vision or end state. In the latest maritime strategy, we listed missions, but the linkage between the missions, the design or architecture of our Fleet, and the resources necessary to build and sustain that Fleet are not clear. That missing coherence led to negative comments on *CS-21*.

Rumelt defines the three elements of a good strategy as diagnosis, guiding policy (which is better defined as strategic design or logic), and coherent actions (see Figure 1). Good strategy is based on a solid diagnosis and provides a coherent plan of action to achieve stated aims. American maritime strategy must do the same. Good strategy must account for a dynamic geopolitical context and interaction with an adversary who has his own goals and options.

It is not locked into a static paper, it evolves and adapts to changing circumstances and an unpredictable world. Thus, we have to understand that our crystal balls are opaque, that critical assumptions must be retested, and planners recognize when new information disproves the basis of the plan or alters crucial decision points. We must constantly be prepared to question received wisdom, evaluate both explicit and implicit assumptions, ruthlessly assess results, and be willing to generate new hypotheses, plans, and solutions. Most of all, we need to ensure the thread or coherence between problem and solution. This, rather than slavish devotion to a glossy document, is real strategy and the best antidote against strategic paralysis.

Figure 1: Elements of Good Strategy

1. A DIAGNOSIS
An explanation of the nature of the challenge. A good diagnosis simplifies the often overwhelming complexity of reality by identifying certain aspects of the situation as being the critical ones.

2. A GUIDING POLICY
An overall approach chosen to cope with or overcome the obstacles identified in the diagnosis.

3. COHERENT ACTION
Steps that are coordinated with one another to support the accomplishment of the guiding policy.

The Purpose of a Maritime Strategy

We lack a common agreement among the naval-strategy community about what a maritime strategy is and what it should comprise. Moreover, we have varying expectations about what the maritime strategy should achieve. Some contend it is enough to articulate a compelling narrative about the purpose of a Navy, others that we must detail a strategic vision, and still others want to argue aggressively for a larger Navy.[8] But as Rumelt contends, "A strategy is not what you wish would happen. It is a set of practical actions for moving forward. It is not a 'dog's dinner' of all the things various parties would like to see done." Our strategy should produce "a focusing of energy and resources on a few key objectives whose accomplishment will make a real difference."[9] There are at least five potential benefits from a comprehensive maritime strategy:

- Defines a shared vision of institutional purpose or mission for its members
- Creates an awareness and consensus on core challenges in a dynamic and competitive environment
- Identifies the ways/means logic to create and sustain a competitive advantage relative to the core challenge. This generates a clear and integrated strategic logic; a coherence between the ends, ways and means of the strategy
- Guides the development and sustainment of maritime capabilities and required capacity (Fleet design and size)
- Generates innovative trans-domain synergies (surface, subsurface, aviation, cyberspace) in the design and application of maritime power

CS-21 arguably achieved the first two of these possible purposes. The maritime leadership needs to decide how far down in descending order they want the next strategy to evolve. The third and fourth objectives would require a more substantial rewriting and redesign of the strategy. There are numerous arguments for and against linking the strategy to an explicit force design. Robert

Kaplan, a senior fellow at the Center for a New American Security, once commented that *CS-21* was subtle, what he called a nuanced call for a larger Navy.[10] The next iteration should be more forceful, and less nuanced.

Strategic Pitfalls or Sins

Rumelt identifies a number of common strategy "sins."[11] The first seven listed in Figure 2 are his. The eighth and ninth are this author's own additions.

1. The essence of strategy is not only about making choices, it is about making choices concerning resources that are almost always constrained. The notion that one writes a strategy and then costs it out for submission as a budget is quite the rage on Capitol Hill, but it does not square with anything to do with strategy. Constraints and restraints impinge on all strategies. Real strategies are ultimately resource-constrained, and our fiscal situation only makes it more obvious today.

2. Rumelt is equally critical of leaders who confuse goals and audacious objectives with strategy. Increasing sales by 200 percent in one year or building a 450-ship Navy may galvanize the audience with one's sheer audacity, but offers little in terms of real strategy unless it can be linked to actionable steps matched with resources. Real strategy or good strategy is about integrated and implementable steps that show how a goal can be achieved and how resources, plans, and policies are coherently linked to progress. Reaching for the stars is commendable, but it is not strategic thinking.

3. The foundation of a strategy is its diagnosis of the strategic environment and core challenges. Rumelt finds that many leaders overlook this phase, and furthermore do not recognize nor state the true strategic problem to be resolved. We probably do not suffer from this problem today, thanks to China's crass assertiveness of late, but the last maritime strategy sidestepped the China challenge to focus on building and sustaining a global international system and access to the global commons, which represents its maritime-based trading distribution

Figure 2: Common Strategy "Sins"

① Failure to recognize or take seriously that resources are scarce.
② Mistaking strategic goals for strategy.
③ Failure to recognize or state the strategic problem.
④ Choosing unattainable or poor strategic goals.
⑤ Not defining the strategic challenge competitively.
⑥ Making false presumptions about one's own competence or the causal linkages between one's strategy and one's goals.
⑦ Insufficient focus on strategy due to trying to satisfy too many different stakeholders or bureaucratic processes.
⑧ Not recognizing that implementation is part of the process.
⑨ Falling in love with the strategy—ignoring environmental change and the fact that strategy is an iterative process not a product.

network. Any *CS-21* writing team will have to address the range of challenges facing the nation and identify the critical issue that the maritime services can address.

4. *CS-21* cannot be accused of being guilty of the sin many corporations commit—choosing grand, even audacious, goals that are hopelessly out of reach. Too many companies seek to double sales in a year, or attempt to grab the lion's share of their market, even though they lack plans, resources, or means of obtaining their goal. Since *CS-21* was more declaratory and explanatory in nature, it contains no explicit goals. However, there was a general understanding that the CNO's oft-expressed plan for a 313-ship Fleet was the implicit goal of the strategy and the Department of the Navy. The next strategy should be more explicit about both the capabilities needed to be acquired and the necessary capacity to underwrite it. The strategy should have some stretch in it—an aspiration (300-plus ships or 12 ships a year) not a delusion (a fleet of 346 and beyond).

5. Rumelt's claim that strategies should be phrased and framed in competitive terms is debatable. One of the strengths of *CS-21* was that it was opportunity-based and not founded on a competitor or threat. Other critics wanted the strategy to identify China as the rising challenge and use that threat as our operational focus and to gain additional resources. This aspect of the strategy drew the most comments.[12] But *CS-21* took a more indirect approach about China and the stability of the international system. Thus, it was based on a strategy of positive aims—a legitimate and more compelling strategy for future partners. This approach defends the extant rules of the road and draws in potential partners interested in maintaining the status quo and willing to work collaboratively toward that. Rumelt's guidance suggests that we reframe the next strategy like War Plan Orange or the *Maritime Strategy* of the 1980s, which was very competitive and operationally oriented. This issue will remain the most contentious component of the updated strategy.

6. The sixth common sin is based on complacency or false assessments about an organization's strengths and weaknesses relative to the operating environment. *CS-21* was balanced about the Navy's need to preserve its global reach and the readiness of our warfighting capability. It was also reasonable about the need to improve in areas like irregular warfare and expeditionary operations. Where the strategy did break down was in what Rumelt insists is the core issue—its coherence or the causal linkages between one's goals and the strategy. Because it was determined that subsequent internal planning documents, such as the classified Navy Strategic Plan, would be where explicit goals, programming priorities and budgets would be detailed, *CS-21* was not able to lay out a coherent Fleet design or even a force structure goal that was causally linked to the mission list or implicit priorities of the strategy. If the revision of *CS-21* does anything, it should help its audience understand this linkage, a coherent and explicit tie between the strategy, its aims, and the means required. Without this fix, any new document is going to face strong criticism.

7. The authors of *CS-21* might plead guilty to this sin, given the excessive number of stakeholders and audiences they tried to satisfy. Who is the ultimate audience for the next issue, the maritime services that must embrace and operationalize the strategy? Is it the Executive branch, Department of Defense bureaucrats, or Congress that ultimately is the arbiter of funding? The current *CS-21* spoke to many constituencies, including future partners and potential adversaries. The next iteration will be read by all, but it must speak directly to its most important audience and not just reflect a bureaucratic lowest-common-denominator. Surely, all these will eventually read the document, and one might want to be aware of the potential misinterpretations that might be made, but ultimately the real audience for this document is the U.S. Navy and its maritime partners who must believe in it.

Two Additional Pitfalls or Sins

8. Conversely, the signatories of *CS-21* did recognize that implementation is part of the strategy-making process. The writing team and the maritime leadership included numerous implementation imperatives—but they did not go far enough since they assumed (erroneously) that the budgeting process was the proper place for execution details. Several major naval initiatives over the past few years, including ballistic-missile defense in the Persian Gulf and Europe, the stand-up of the 10th Fleet and U.S. Marine Cyber Command, and the enormous impact of funding the Navy's contributions to the nation's strategic deterrent (a potential $100 billion development and acquisition bill) are not addressed in the strategy. We have published a strategy, but arguably not truly implemented it.

9. Some Cold War strategists can be accused of mortal rigidity about containment, but it was flexible in its application. So too for *CS-21*. But clear-eyed strategists must always scan the environment for ongoing changes and even disconfirming evidence about geostrategic dynamics. A good strategist recognizes that assumptions are not written in stone,

and that strategy is really an iterative and continuously renewable process.[13] It is not about writing a glossy document—it's about constantly adapting to new circumstances.

Ultimately, crafting the strategy is only the beginning of a journey—good strategy is not just a product but a process of execution and adaptation. "Like a vessel under sail, grand strategy is at the mercy of uncontrollable and often unpredictable political, economic, and military winds and currents."[14] Naval strategists must be alert to changes in context, take a sextant bearing, and apply constant tiller correction.

"An Urgent Matter"

In conclusion, reversing the decline in U.S. strategic competence is recognized as an urgent matter.[15] All too often in the United States there is "an unrecognized black hole wherein strategy should have resided," what the Anglo-American strategist Colin Gray titled our Strategy Deficiency Syndrome.[16] In the past, we could outspend, outproduce, and outfight our adversaries and overcome. No more—it's a cliché, but it is time to out-think our opponents as well.

The naval strategy community is not as guilty as other components of our national security bureaucracy, but we can do better. A refresh to *CS-21*, even a major rewrite, is a healthy thing. There are critical aspects of the strategic environment, including our diminished fiscal foundation, to consider. Given the recent guidance issued by Secretary of Defense Leon Panetta, a refreshed strategy is entirely appropriate.[17] It also should reflect Chairman of the Joint Chiefs of Staff General Martin Dempsey's direction for Joint Force 2020.

Good strategy involves hard choices, clear objectives, a continuous assessment of risks, and priorities. We have problems with the latter far too frequently. As the retired Army strategist Rick Sinnreich noted, "A vital requirement of successful strategic design is to bound the universe of objectives, recognizing that the desirable is not the same as important, nor the important the same as urgent."[18] A refreshed iteration of *CS-21* must do the same.

The CNO's initiative to update our strategy is applauded, as our maritime services must adapt to real-world events and act on behalf of U.S. security interests in dangerous waters. Our emerging strategy is the next step in the adaptation to an ever-evolving context in a dynamic world. However, we need a "good" strategy, not a vision or ephemeral restatement of the obvious. The greater the challenge, the more a good strategy focuses and coordinates efforts to achieve a powerful competitive punch and effect.[19] Our nation needs the potent punch and stabilizing effect of a powerful Fleet.

Notes

1. Hew Strachan, "The Lost Meaning of Strategy," *Survival* (Autumn 2005): 33–54; Aaron Friedberg, "Strengthening U.S. Strategic Planning," *The Washington Quarterly*, 31:1, 47–60.
2. CDR Michael Junge, USN, "So Much Strategy, So Little Strategic Direction," U.S. Naval Institute *Proceedings* (February 2012): 46–50.
3. Williamson Murray, MacGregor Knox, and Alvin Bernstein, eds., *The Making of Strategy: States, Rulers, and War* (New York: Cambridge University Press, 1994), 1.
4. Henry Mintzberg, *The Rise and Fall of Strategic Planning* (New York: Free Press, 1994), 318–21.
5. Henry Mintzberg, Joseph Lampel, and Bruce Ashland, *Strategy Safari: A Guided Tour Through the Wilds of Strategic Management* (New York: Free Press, 1998).
6. Richard Rumelt, *Good Strategy, Bad Strategy, The Difference and Why It Matters* (New York: Crown Business, 2011), 20.
7. Richard Rumelt, "The Perils of Bad Strategy," *McKinsey Quarterly* (June 2011): 11.
8. See William T. Pendley, "The New Maritime Strategy: A Lost Opportunity," *Naval War College Review* (Spring 2008); countered by Robert C. Rubel, "The New Maritime Strategy, the Rest of the Story," *Naval War College Review* (Spring 2008).
9. Richard Rumelt, "Good Strategy/Bad Strategy On Sale," *Rumelt's Web Journal*, July 28, 2011, www.strategyland.com/2011/gsb-on-sale/.
10. Robert D. Kaplan, "America's Elegant Decline," *The Atlantic*, November 2007, www. theatlantic.com/magazine/archive/2007/11/america-8217-s-elegant-decline/6344/.

11. See Andrew F. Krepinevich Jr. and Barry D. Watts, *Regaining Strategic Competence* (Washington, DC: Center for Strategic and Budgetary Assessments, 2009), 33–4.
12. See Robert O. Work and Jan van Tol, "A Cooperative Strategy for 21st Century Seapower: An Assessment" (Washington, DC: Center for Strategic and Budgetary Assessment, March 26, 2008); Frank Hoffman, *From Preponderance to Partnership, American Maritime Power in the 21st Century* (Washington, DC: Center for a New American Security, 2008).
13. See Williamson Murray, *War, Strategy and Military Effectiveness* (New York: Cambridge University Press, 2010), 136–7, 164–6.
14. Ibid., 256.
15. Barry D. Watts, *U.S. Combat Training Operational Art, and Strategy Competence, Problems and Opportunities* (Washington, DC: Center for Strategic and Budgetary Assessments, 2008).
16. Colin S. Gray, *The Strategy Bridge, Theory for Practice* (New York: Oxford University Press, 2010), 247.
17. Leon Panetta, *Sustaining U.S. Global Leadership: Priorities for 21st Century Defense* (Washington, DC: Department of Defense, January 5, 2012).
18. Richard Sinnreich, "Patterns of Grand Strategy," in Williamson Murray, Richard Hart Sinnreich, James Lacey, eds., *The Shaping of Grand Strategy: Policy, Diplomacy, and War* (New York: Cambridge University Press, 2011), 263.
19. Rumelt, *Good Strategy/Bad Strategy*, 4.

6 "THE GLOBAL MARITIME COALITION"

LCDR James Stavridis, USN

At the earlier-mentioned "Maritime Security Dialogue" conference in October 2014, sponsored by the Center for Strategic and International Studies, Chief of Naval Operations Jonathan Greenert concluded his opening remarks by referring to this article by then-lieutenant commander James Stavridis, pointing out that now-retired admiral Stavridis was "a person ahead of [his] time."

In this forward-thinking article Commander Stavridis acknowledges that the "contributions of the free world are significant" and urges U.S. planners to "develop contingency plans that consider these contributions."

"THE GLOBAL MARITIME COALITION"
By LCDR James Stavridis, USN, U.S. Naval Institute *Proceedings* (April 1985): 58–64.

The contribution of Free World navies to U.S. maritime strategy is often overlooked or understated by U.S. planners. Too often, assessments of allied navies are limited to hardware profiles of specific ship capabilities, personnel and training levels, and other detailed statistics. This information is important, but the broader strategic picture must also be formulated. Free World naval planners must work to develop a coordinated "maritime coalition" strategy. U.S. planners must assess the capabilities and limitations of Free World navies in the largest sense, because ". . . our greatest remaining strategic advantage over the U.S.S.R. is that we have many rich allies whereas it has only a handful of poor ones."[1]

Global Maritime Coalition

In assessing the contributions of the Free World navies to Western defense, it is possible to think of these forces as being linked by an informal international network that might be called the "global maritime coalition." Though loosely structured, it does have many formal associated linkages, notably the mutual defense treaties of the United States and immediate allies, such as NATO, SEATO, the U.S.–Japanese treaties, the Rio Pact, and so forth. The navies involved operate together in exercises such as RimPac, Unitas, Team Spirit, and others. Port visits are often exchanged between regional partners, and deployment visits to other member countries are often undertaken.

Obviously, there is no formal governing mechanism, and disputes between members (such as Britain and Argentina) are not infrequent. The members often take opposing positions in international maritime fora, such as the Law of the Sea Treaty discussions or International Maritime Organization (IMO) deliberations. Yet all are generally held together by a shared desire to foster a policy of freedom on the high seas, maintain strong defensive forces to deter aggression, and keep sea lines of communication (SLOCs) open. Figure 1 depicts the general outline of the global maritime coalition.

Much of the direction of the structure is provided by the United States, via multilateral and bilateral negotiations and treaties. Overall, however, the global maritime coalition is more a state of mind on the part of Western planners than anything else. There are two primary purposes for the coalition. First, the most important, the concept allows Western planners a strategic framework to consider in planning contingencies to respond to Soviet aggression or action by radical developing countries (Iran, Iraq, Libya, Vietnam, North Korea, Syria,

Figure 1: The Global Maritime Coalition

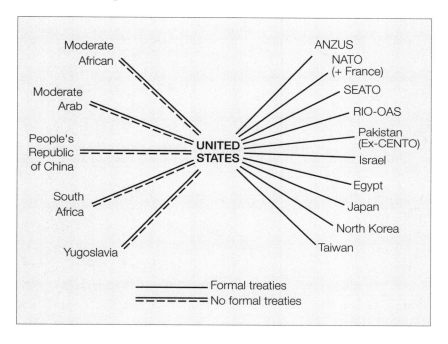

etc.). Second, the coalition provides a means for U.S. planners to assess readiness and work toward integrating its members into a more cohesive and better prepared operating force.

Geographical-Naval Contributions

Perhaps the single greatest contribution of the Free World navies in the broad strategic sense is their geographic position. Although the United States operates at the center of this informal global maritime coalition, it is a geographically isolated island-state. Likely areas of conflict, whether against the Soviet Union or radical developing states, are generally a considerable distance from the United States. Although the forward-deployed naval strategy is designed to allow U.S. involvement at great distances, the Free World navies are critical in many situations, either as a first line of response/defense or as reinforcements.

As the "Navy League's 1984–85 Resolutions" point out, "The Soviets have gained foothold access to the world's shipping choke points—the Suez Canal, Southeast Asia, Panama, and Southern Africa—and have established military strength in positions where they could deny free world access to critical resources in the Persian Gulf, Africa, and Southeast Asia.[2] Beyond the looming geographical threat of the Soviet Union, many radical developing states are capable of holding U.S. interests at risk in the global periphery. Only credible allied naval forces operating as part of a global maritime coalition will be capable of responding to a wide diversity of simultaneous threats. A number of examples illustrating this fact follow.

Europe: From a geographical standpoint, the forces of the global maritime coalition in the European NATO regions will be important in opposing any Soviet invasion on the continent. Even in a cold war situation, "they provide a geographical arrangement of the greatest importance for surveillance forces."[3] This surveillance function includes underwater acoustic listening stations, high-frequency/direction-finding sites, bases for their own and U.S. surveillance aircraft and ships, and radar sites. In any threat scenario, such early warning will be essential to solid mutual defense. Beyond surveillance, the on-station capabilities of European NATO forces will be significant in any U.S.–Soviet confrontation. Of particular importance will be the following:

- Antisubmarine warfare (ASW) efforts in the Greenland–Iceland–United Kingdom Gap, the North/Norwegian seas, the Baltic Sea, and the Danish Straits, as well as the Sea of Marmara, the Black Sea, and the eastern Mediterranean choke points to the south. These ASW efforts will include long-range maritime patrol aircraft and surface ship efforts, as well as some ASW submarine efforts. Most of the European NATO countries have such forces.
- Antisurface warfare (ASUW) operations to attrite the Soviet naval forces as they sortie, conducted by high-speed patrol craft firing antiship cruise missiles, diesel submarines in an ASUW role, and land-based air. The forces of Germany in particular, with the *Bremen*-class frigates,

"Type-206" coastal submarines, and "Type-143A" fast attack craft, will be important. Norway's northern frontier, of course, is located only 80 nautical miles from the Soviet Northern Fleet bases. Much ASUW action might be based in that region.
- Minelaying and mine clearance operations in both the northern and southern theaters will fall largely to allied European NATO naval forces. Belgium's bases, including Ostende, will be critical, and the lowland navies will participate in such operations effectively.
- Support for the European land flanks may be conducted with European NATO forces, particularly if U.S. carrier battle groups are occupied in other strategic missions. The U.K. and French forces can provide such supply support, naval gunfire support, air strikes, and jamming.
- Atlantic operations will be aided considerably by the use of Portuguese forces. Geographically, Portugal will be of immense importance to the maritime coalition. This results from the importance of Lisbon as a port, as well as the Azores and Madeira island bases.
- In the eastern Mediterranean, Greece and Turkey will hopefully be able to lay aside their differences and face a mutual defense problem consisting of possible land invasion from their northern borders and a sortie of Soviet forces into the Mediterranean. Their contributions to the maritime coalition will be significant given their position astride the choke points of the eastern Mediterranean. With eight modern submarines and numerous missile boats, Greece and Turkey can attrite Soviet forces in the region with considerable effect.

Middle East: The presence of various Free World naval forces in the Persian Gulf, Suez, and Red Sea areas will be geographically critical to maritime defense for many scenarios. The small but growing naval forces operated by the Saudis and other gulf states include excellent missile-firing platforms such as the French-built "F-2000" frigates, *Badr* corvettes, and *Al Siddig* fast attack craft.[4] The Egyptian and Israeli navies might work together in missile operations, limited ASW, and mine clearance/laying in the Suez–Red Sea waterway. The Omani Navy is growing in capability under U.S. and British guidance and support.

Together, these forces might serve as a counterweight to radical developing state naval forces (Iran, Iraq) in the area if U.S. and other Western assets were employed elsewhere.

Indian Ocean: Often overlooked in the Indian Ocean region are the significant assets maintained by the French. With a flotilla and frequent rotations of aircraft carriers into the region, the French have a significant naval presence. Their geopolitical capabilities, given such former French colonies as Reunion, Madagascar, and the Ascension Islands could be important. In addition, Thailand and Malaysia both operate small but capable naval forces in the eastern Indian Ocean that might serve as a counterweight to possible anti-U.S. action by India or the Soviet Union.

Pacific Ocean: From a geographical standpoint, the United States is blessed with well-placed allies in the Pacific region. Seated in the heart of the western Pacific, the Philippine Islands provide bases that are even more important since the fall of Vietnam and the pending return of Hong Kong to the Chinese. The Japanese geographical position, blocking the sorties of Soviet forces from Pacific bases, is likewise critical, and the capable Japanese Maritime Self-Defense Force is preparing for expanding missions in SLOC protection, attrition of Soviet forces, and ASW operations in local waters. Such platforms as the *Hatsuyuki* guided-missile destroyer and *Yushio*-class submarines will be important. The South Korean Navy, though focused on North Korea, can be a significant asset in western Pacific operations.

The northern Pacific, a region of increasing strategic attention for U.S. planners, will require geographical support from Japanese and Korean locations. The southern Pacific also includes several maritime coalition members of some importance. Small but capable, the Australian and New Zealand navies are important players in many Pacific operations and in SLOC defense.

South America: The South American forces operate under the Rio Pact alliance, but they also operate as part of the maritime coalition. The Unitas cruises are indicative of the ongoing U.S. participation and interest in naval affairs in the region. As the threat from radical developing states in the region (Cuba and Nicaragua) grows, the value of such regional powers as Brazil and Venezuela will continue to grow. The Brazilian *Niteroi*-class frigates (produced jointly with the United Kingdom) and the Venezuelan "Lupo"-class frigates (Italian construction) will be important considerations for U.S. planners in this volatile region. In many scenarios, especially where U.S. carrier battle groups must be involved in European or Southwest Asian defense operations, the South American naval forces might be critical to maintaining South Atlantic SLOCs and keeping order in the Caribbean basin.

Indeed, a *New York Times* article stated that the U.S. Navy is conducting discussions about "operating together to protect sea lanes of communication, watch for submarines, and intercept infiltrators along coastlines."[5] Both Venezuela and Colombia were mentioned specifically in the article. The Chief of Naval Operations will host a conference in Norfolk, Virginia, in May 1985, "where Latin navy chiefs will be asked to discuss mutual support, communications and threat warnings."[6]

Africa: Although the growing controversy of any country developing close relations with South Africa (because of its apartheid policy) precludes its full participation in the global maritime coalition, it remains a fact that the country sits on a critical global choke point. South Africa continues to expand its naval forces. As a counterweight to radical developing countries on the central-southern African plains, however, South African forces might make a *de facto* contribution to U.S. interests. The growing Soviet basing structure on the African continent (particularly in West Africa) may be balanced to some degree by the South African presence.

Both with forces on station and in providing critical forward basing rights, the Free World navies will make valuable contributions to the global maritime coalition. Forward-based allies will likely be a necessity if the United States is to be prepared to conduct capable forward maritime defense in the global periphery. These allies can provide war-fighting forces as a first line of defense or in a critical reinforcement role, sites for forward storage of supplies, ammunition, and petroleum, oil, and lubricants, repair facilities, and recreation and relaxation ports.

Geographically, the importance of such U.S. naval complexes as Subic Bay (at the hub of the western Pacific), Diego Garcia (in the southern Indian Ocean), Naples (in the heart of the Mediterranean), Bahrain (in the Persian/Arabian gulfs) cannot be understated. In the face of continued Soviet expansion to new bases in Southeast Asia, Western Africa, Central America, and the Pacific, the importance of these installations will continue to be critical.

Political-Economic Contributions

The Free World naval forces also make a political-economic contribution to the global maritime coalition. Although space does not allow an in-depth analysis of the total political or economic ramifications of the Free World naval contributions, some of the most important and more tangible follow.

Intelligence-Gathering and Sharing: Early warning and intelligence-gathering have significant political ramifications. The need to gather intelligence in countries difficult for U.S. agencies to penetrate (because of cultural differences) is obvious. Much of the information provided by the Israelis, for example, is enlightening and useful and would not be available through other channels. U.S. joint-sharing of intelligence and information with other members of the global maritime coalition helps keep the coalition working together.

Joint Political Operations: The best recent example of joint naval operations conducted for political reasons is the Suez mine clearance operations that followed the terrorist mining operation. The United States, France, the United Kingdom, and other global maritime coalition members all participated in a demonstration that not only served to remove the mines but also made a strong political statement about the West's resolve to keep SLOCs from the Persian-Arabian Gulf open to oil and other merchant traffic. Another ongoing political example is the annual RimPac exercise conducted in the Pacific between units from the U.S., Australian, New Zealand, Japanese, South Korean, and other allied naval vessels. Likewise, the Unitas exercises make a political statement in addition to providing training for the crews involved.

Exchanges and Training: In addition to the macro-level training conducted with units at sea, another valuable contribution of the Free World navies is that of individual exchanges and training. Many foreign officers and men visit the United States to study at a wide variety of Navy schools and training facilities. The United States, in turn, sends officers to a variety of billets throughout the Free World annually. Their observations and interaction with foreign naval personnel become informally incorporated into U.S. naval strategy in a very effective manner.

International Fora: The interaction between naval personnel involved in international fora can often lead to surprising results. Informal negotiations in the Law of the Sea Conference and the International Maritime Organization are examples.

Economics: The United States participates in a great deal of international trade and exchange that directly contributes to its economic well being and strong defense industries. America's partners in the global maritime coalition are also America's strongest trading partners. Indeed, the economic factor is perhaps the single largest common denominator among the members of the coalition, and the need to keep trade moving across free sea-lanes is a driving factor in policy and strategy.

Overall, the intelligence, intellectual exchange, political presence missions, joint exercises, and economic interaction are all important contributors to the overall maritime defense of the Free World.

Recommendations

Several recommendations for U.S. strategic planning can be offered that might enhance the potential for positive interaction in the global maritime coalition. These recommendations are stated from a U.S. point of view, optimizing outcomes and relationships that favor U.S. security.

Planning: U.S. national-level planning should consider the contributions of the Free World navies, in both a current and projected sense. This should be done by analyzing such contributions from geographic-naval and political-economic standpoints. The long-range goal should be to shift burdens, where appropriate, to alliance systems within the global maritime coalition, while strengthening existing relationships.

Linkage Development: The global maritime coalition consists of formal treaties, bilateral accords, statements of friendship, trade relationships, and so forth. Every effort should be made from a U.S. standpoint to develop maritime linkages with other Free World maritime-oriented states. This should be done as part of a coherent national ocean policy with a well-defined component of naval strategy. Some methods of doing this might include:

- Continually emphasize joint operations with various regional partners within the global maritime coalition.
- Sponsor joint building projects like naval weaponry and research and development. Mutual projects will eventually lead to better mutual support in coalition operations.
- Develop exchange programs, not only for individuals, but perhaps for units, giving maximum operations interaction between the United States and other maritime coalition partners.
- Sponsor nonmilitary ocean projects, involving fishing, deep seabed mineral recovery, ocean energy projects, and underwater habitat research. These could be done in a variety of regions, with beneficial flow between segments of the coalition. The research would eventually be compatible with military projects.

Conferences: The concept of U.S.–sponsored maritime conferences to discuss ocean issues among members of the coalition opens a variety of possibilities. These conferences could be attended by naval officers (possibly chiefs of naval operations), appropriate civilian officers from Department of State equivalencies, and ocean experts. Both naval-military and civilian-oriented conferences could be developed.

Technology Transfer: The sharing of appropriate technology between coalition partners, with the necessary supervision and control, would eventually lead to better coalition maritime defense with more compatible and capable coalition-wide systems.

Operations: Increased operating time between coalition members would improve U.S. and international defense against possible threats. The RimPac exercises in the Pacific are a good example of the potential for such operations. The continued development of English as the *lingua nautical franca* for coalition operations would be enhanced, operators would obtain training, and war gaming could be verified.

Conclusions

The contributions of the Free World navies are significant. As part of a loosely structured global maritime coalition, such forces make important geographical-naval and political-economic contributions to international peace and defense. U.S. planners should continue to develop contingency plans that consider these contributions. At both a political and a military level, the United States must enhance the structure of the global maritime coalition. Only by doing so can the United States help create an atmosphere which maximizes the capabilities and potential of the allied forces.

Notes

1. Robert Komer as quoted in Michael McGwire, Book Review, *Proceedings* (July 1984): 118.
2. "Navy League's 1984–85 Resolutions," *Sea Power* (September 1984): 46–47.
3. Foreword to *Jane's Fighting Ships 1984–85* (New York: Jane's Publishing Inc., 1984), as quoted in *Sea Power* (September 1984): 33.
4. "Saudi Arabia," *Conway's All the World's Fighting Ships 1947–1982, Part II* (Annapolis, MD: Naval Institute Press, 1984), 446.
5. Richard Halloran, "Latin Naval Help is Sought by U.S.," *New York Times,* October 21, 1984, 6.
6. Ibid.

7 "WHY 1914 STILL MATTERS"

Norman Friedman

Prolific (putting it mildly) author Norman Friedman derived this article from his book *Fighting the Great War at Sea: Strategy, Tactics, and Technology* (Naval Institute Press, 2014). While acknowledging that "history never repeats," Dr. Friedman correctly contends that "it is often instructive to look at the mistakes of the past." This look back is well taken as he recalls the decisions (and consequences) made during World War I. When one thinks of this horrific conflict, images of trench warfare often dominate, but Dr. Freidman points out that this was very much a maritime war, providing much food for thought for modern naval strategists.

"WHY 1914 STILL MATTERS"

By Norman Friedman, U.S. Naval Institute *Proceedings* (August 2014): 62–69.

Today, as a century ago, the fact that war between trading nations would be ruinous does not necessarily mean that its outbreak is impossible.

Imagine that your closest trading partner is also your most threatening potential enemy. Imagine, too, that this partner is building a large navy specifically targeted at yours, hence at the overseas trade vital to you. Does that sound like the current U.S. situation with respect to China? It was certainly the British situation relative to Germany a century ago, on the eve of World War I. History never repeats, but it is often instructive to look at the mistakes of the past. The worse the mistakes, the more instructive. No one looking at the outbreak and then the course of World War I can see it as anything but a huge mistake. Hopefully we can do better.

The worst mistake, from a British point of view, was to forget that this was a maritime war. Had the British not entered the war at all, it would have been a European land war. Once Britain entered, the character of the war changed, not only because Britain was the world's dominant sea power, but also because the British Empire—including vital informal elements—was a seaborne entity, drawing much of its strength from overseas. As an island, Britain was almost impossible to invade. Centuries earlier, Sir Francis Bacon had written that he who controls the sea can take as much or as little of the war as he likes. The sea power did not have to place a mass army ashore. That was not necessarily its appropriate contribution to a coalition effort.

Our memory of World War I overwhelmingly emphasizes the blood and horror of the Western Front, to which the U.S. Army and Marines were assigned when entering the conflict in 1917. The war at sea is usually dismissed as a sideshow, at best an enabler for the more important action ashore. That view obscures the reality that the war was shaped by maritime considerations, and, at least as importantly, the potential that seaborne mobility offered the British and the Allies. The one instance of a strategic attack from the sea, Gallipoli (the Dardanelles campaign), is usually dismissed as an attempt by First Lord of the Admiralty Winston Churchill to gain publicity for the Royal Navy. In fact, it was a high-risk, high-payoff operation supported by the British cabinet for very rational reasons. That it failed does not make it a foolish bit of grandstanding. It only proves that planning and execution were extraordinarily poor. Our memory of how the war was fought obscures the fact that there were real alternatives, at least for the British.

Our present situation is more like that of the British than that of their continental allies. How well would we do in a similar situation? We were actually confronted by one during the Cold War. The U.S. Navy's Maritime Strategy was an alternative way to fight a continental war. It is still worth thinking about.

The Accidental Army

When the British entered World War I, Prime Minister Herbert Asquith expected the French and the Russians to provide the bulk of the forces on land; the British army's contribution in France was to be largely symbolic.[1] The British expected the French to hold the German army in the west while the "Russian steamroller" smashed from the east. However, Asquith casually approved War Minister Lord Herbert Kitchener's program to create massive "New Armies" (without ever being forced to explain their rationale). The British slid into creating the largest army in their history. Once that army existed, it could not be denied to the French when they found themselves in serious trouble in 1915. Once there, it could not easily be withdrawn. Most of the 800,000 British Empire troops killed in World War I died on the Western Front.

Were these horrific losses inevitable? Given the sheer depth of modern economies and the power of the defense, the war on land would surely have been a protracted bloodbath. Did it have to be a *British* bloodbath? Asquith was Prime Minister of the United Kingdom, not of some Franco-British combination. It was clearly in the interest of the French that the British army fought alongside theirs and helped preserve France. Was that in British interests, too? How deep should coalition partnership cut? Could the British have fought a more maritime war? In Vietnam, in Iraq, and in Afghanistan the United States has faced the question of how far to go in support of a coalition partner.

Perhaps the saddest feature of British prewar and wartime planning was Admiral Sir John Fisher's futile attempt to point out that although (as everyone agreed) no success on the Western Front could be decisive, the Germans were extraordinarily sensitive to threats to their Baltic coast—a place accessible by sea, albeit with considerable danger. Unfortunately, Fisher made his point, both before and during the war, in an obscure, even mystical way.[2] The often-denigrated

Dardanelles operation was a remnant of the abortive British maritime strategy; it was intended to help sustain Russia. Fisher's great objection was that it would swallow forces he thought could have been used more effectively in the Baltic—again, to support the Russians on what he and others thought was the decisive front.

The deeper reason for British planning failure is that almost up to the declaration of war virtually no one in London believed that there could ever be a war. It was widely accepted that, because the major economies were so closely intertwined, any war would be disastrous. The Britain of 1914 was a much more modern nation than its European partners. International finance played a larger part in the British economy than in any other. The financial sector still considers war futile: If one asks someone on Wall Street right now whether a war with China is possible, the answer is emphatically no, that would be ruinous. If the point of government is to maintain national prosperity, big wars are absurd. The British government of the years before 1914 did not, it seems, understand that those governing Germany had rather different ideas. How well do we understand how foreign governments think? Are big wars really obsolete?

Economy as Weapon = Double-Edged Sword

In effect, those in London thought that what was much later called mutual assured destruction prevailed. War fighting and therefore war planning were of little account. The British army commitment to France was much more symbolic than real, an attempt to show the French that the British would back them in the event of a crisis. This plan was accepted (though not, it seems, wholeheartedly) largely because it was far more important that prewar War Minister Richard Haldane led an influential faction in the governing Liberal Party than that the army's favored plan for deployment in France made much military sense.

The British government naturally became interested in economic attack as a means of quickly concluding any war that broke out. The Admiralty became an advocate of such warfare as a natural extension of the traditional naval economic weapon of blockade. In 1908 a prominent British economist pointed out that in a crisis the British banks, which were central to the world economic system, could attack German credit with devastating results.³ Somewhat later the British banks pointed out that since Germany was Britain's most important trading partner, any damage would go both ways. Banking had to be omitted from the arsenal of economic weapons. It turned out that sanctions imposed on Britain's main trading partner were less than popular in the United Kingdom—and that they badly damaged the British economy which depended on trade. For example, a prohibition against trading with the enemy made it necessary to prove that every transaction was *not* with the enemy. It was not at all clear that the damage done to the British economy did not exceed that done to the German.

In pre-1914 Europe the single life-and-death problem for most governments was internal stability. Most thought in domestic terms. For example, the British Liberal Party resisted naval and military spending because it considered social spending vital for British stability. The tsarist government in Russia sought to create a strong peasant class as a bulwark against socialist workers (assuring grain exports, which would create the prosperous peasant class, required free access to the world grain market via the Dardanelles). However, the Austro-Hungarian government feared nationalist upheaval triggered from outside, most notably from Serbia (and was unable to promote internal reform).

German leaders thought they faced an imminent internal crisis.[4] The perceived crisis was the rise of a hostile majority in the Reichstag, the lower house of the German parliament. Although hardly comparable to the British Parliament, the Reichstag was responsible for the budget. In elections from 1890 on, the Social Democrats, whom the Kaiser and his associates considered dangerous revolutionaries, consistently won majorities of the vote, but because seats were gerrymandered they did not win a majority in the Reichstag until 1912. The German army's general staff considered itself and the army the bulwark of the regime. Although in theory the Kaiser ruled Germany, in fact he had been sidelined for several years. Army expansion, which might be associated with the sense of internal crisis, began in 1912.

The following year the nightmare became visible, as the Reichstag passed a vote of no confidence after the army exonerated an officer who had attacked a civilian in Alsace.[5] The vote did not bring down the government, because Prime

Minister Theobald von Bethmann-Hollweg was responsible to the Kaiser rather than to the Reichstag. The center-left coalition shrank from rejecting the year's budget. However, there was a sense of escalating internal crisis. A member of the German general staff told a senior Foreign Ministry official that his task for the coming year was to foment a world war, and to make it defensive for Germany so that the Reichstag would support the war.[6]

In this light, the event that precipitated the war—the assassination of the Austrian crown prince Franz Ferdinand—seems to have been much more a useful pretext than the reason the world blew up. The Kaiser was largely on the periphery of rapidly unfolding events during the crisis. He kept asking why the army was attacking France when the crisis was about Russia and Serbia. Do we understand who actually rules countries that may be hostile to us?

Internal Motivations, External Aggression

In 1912–14 the German army general staff could look back to 1870. By drawing France into a war at that time, Prussia had created the German Empire. The spoils of that war were a way of showing that it had been worthwhile, but the war was really about the internal political needs of the German state. In 1914, the general staff doubtless expected that victory would shrivel the Social Democrats (a 1907 military victory over the Hottentots in Africa had reversed their rise, though only briefly). No other military seems to have had a record of deliberately instigating war as a specific way of gaining an internal political end. After World War I, there was a general sense that the German general staff had been responsible for the war, but not to the extent that now seems apparent.

At one time a standard explanation for enmity between Britain and Germany, leading to war, was commercial rivalry. It was taken so seriously that interwar U.S. Navy war planners used British–U.S. rivalry to explain why a war might break out between the two countries. Similarly, one might see Chinese–U.S. trade rivalry as a possible cause of war. However, those concerned with commerce are too aware of how ruinous war can be. Wall Street really does prefer commercial competition to blowing apart its rivals. It has too clear an idea of what war might mean. Naval wars connected with commercial rivalry were fought before commercial and financial interests came to dominate governments. The perceived need to keep the state alive is a very different matter, and it seems to have been what propelled Germany in 1914. Do we see similar motives at work now, or in the near future? The lesson of 1914 is that others' decision to fight is far more often about internal politics than about what we may do.

The Vital Importance of Coalitions

British strategy in 1914–15 may not seem odd in itself, but it is decidedly odd in the context of other wars the British fought on the continent. Everyone in the 1914 Cabinet knew something of the Napoleonic Wars, though probably not from a strategic point of view. That was unfortunate, because they might have benefited from seeing the new war in terms of the earlier one. The British fought Napoleon as a member of a coalition. They watched their coalition partners collapse, to the point where they alone resisted Napoleon. They were forced to agree to a peace in 1801, which they rightly considered nothing more than a pause in the war—and they used that peace to consolidate what advantages they could.

Once the war against Napoleon resumed, the British wisely made it their first step to insure against invasion by blocking and then neutralizing the French and their allied fleets. Once they had been freed from the threat of invasion by the victory at Trafalgar, they could mount high-risk, high-gain operations around the periphery of Napoleon's empire. Ultimately that meant Wellington's war on the Iberian Peninsula. Napoleon realized that he could not tolerate British resistance. Since he could not invade, he was forced into riskier and riskier operations intended to crush Britain economically. His disastrous 1812 invasion of Russia was in this category (it was intended to cut off Russian trade with Britain). The British limited their own liability on the continent. Knowing that they could not be invaded (hence defeated), they could afford to be patient—and they won. Victory was a coalition achievement, which is why it did not matter that so many of the troops at Waterloo were not British.

World War I was shaped by the fact that Britain entered it. Until that moment, the German army staff could envisage a quick war which would end in the West with the hoped-for defeat of the French army. Once Britain was in the war, no German victory on land could be complete. Ironically, the Germans guaranteed that Britain would enter the war by building a large fleet specifically directed against it. Some current British historians have asked whether it was really worthwhile for the British of 1914 to have resisted the creation of a unified Europe under German control. They have missed the maritime point. In 1914 the British saw the Germans as a direct threat to their lives, because the Germans had been building their massive fleet. By 1914 most Britons well understood that their country lived or died by its access to the sea and to the resources of the world. The Royal Navy had worked hard for nearly 30 years to bring that message home. It resonated because it was true. In 1914 the British government would have had to fight public opinion to keep the country *out* of a war the Germans started.

The German decision to build a fleet seems, in retrospect, to have been remarkably casual. The fleet was completely disconnected from the war plan created by the army's general staff; it had no initial role whatsoever. The German navy came into its own only when it became clear that the army could not achieve a decision on land. Then it was not so much the big fleet (that had caught British attention prewar) but the U-boats that Admiral Alfred Tirpitz, the fleet's creator, grudgingly built. The British government might well have decided to oppose Germany in 1914 to preserve the balance of power in Europe—a historic British policy—but without the obvious threat of the German fleet its decision would not have enjoyed anything like the same level of support.

In 1939 the British again faced a continental war. Everyone in the British government had experienced World War I as a horrific bloodbath. This time the British consciously limited their liability. It helped that by 1939 they believed that the Germans could not destroy the United Kingdom by air attack (thanks to radar and modern fighters), so that as in World War I, Britain was a defensible island. Winston Churchill, who had a far more strategic viewpoint than most, certainly did not intend to surrender when the British were ejected from the continent in 1940. He understood that the overseas Empire and the overseas world could and would support Britain against Germany (which is why the Battle of the Atlantic was his greatest concern). He also understood that it would take a coalition to destroy Hitler.

During the Cold War, NATO faced a continental threat not entirely unlike that the British had faced in 1914 and in 1939. Attention was focused on the Central Front, unfortunately so named because it was in the center between the alliance's northern and southern flanks. The U.S. Navy offered a maritime alternative, both in the 1950s and in the 1980s. Captain Peter Swartz, U.S. Navy (Retired), who chronicled the U.S. Navy's Maritime Strategy, summarized the way that a maritime power deals with a land power: It combines a coalition with its own land partner and it exploits maritime mobility to cripple the enemy army.

"Hard Thinking About the Object of War"

Not being able to end a war may seem to be a tame sort of disadvantage to the land power sweeping all before it in Europe. However, both in Napoleon's time and during World War I, the land power (France and Germany, respectively) found that it could not stop fighting. Its effort to knock the British out of the war eventually brought in enemies the land power could not handle. In Napoleon's time that was the Russians, whose territory absorbed the French army, and whose limitless mass of troops eventually helped invade France. Obviously there were many other contributions to French defeat, including Wellington's campaign in Spain, but the point is that none of that would have mattered had Napoleon been able to end the war as he liked.

In World War I the Germans found that their only leverage against the British was to attack their overseas source of strength, either at source in the United States or at sea en route to Britain. Either move was risky. Unrestricted submarine warfare against shipping led to angry reactions from the United States; in 1915–16 the German Foreign Ministry convinced the government (i.e., the general staff) to pull back. As an alternative, in 1916 the Germans organized the sabotage of munitions plants supplying the Allies, most notably Black Tom in New York Harbor. Although the U.S. government almost immediately discovered that the Germans had caused the Black Tom explosion, President Woodrow Wilson badly wanted to stay out of the war. That was not enough for the German general staff. Against Foreign Ministry opposition, it turned again in February 1917 to unrestricted submarine warfare as a way of strangling the Allies.

It was understood that resumption of such warfare would probably bring the United States into the war. With this possibility in mind, the Germans authorized their diplomats in Mexico to offer an alliance under which Mexico would regain the territory it had lost to the United States 60 years earlier: California, New Mexico, Arizona, Nevada, and Texas. Revelation of this Zimmermann Telegram helped bring the United States into the war on the Allied side. U.S. naval and industrial resources helped neutralize the German U-boat campaign in the Atlantic. The U.S. Army and Marines Corps tipped the balance of power in Europe, though it was at least as important that the British and the French became adept at all-arms warfare.

It is also possible that, in the end, the Western Front, where so much blood was spilled, was not decisive in itself. In 1918 the defense still enjoyed considerable advantages. The Germans told themselves that they could shore up their defense in the West, but in September and October 1918 their position in the south, the area in which maritime power had made Allied action possible, collapsed. Whatever they could do on the Western Front, the Germans could not spare troops to cover their southern and eastern borders. In this sense the collapse in the south (of Austria-Hungary, Turkey, and Bulgaria) may have been far more important than is generally imagined.

Maritime never meant purely naval. Success came from using land and sea forces in the right combinations. Maritime did demand hard thinking about the object of the war. In 1914, was it to preserve France or above all to defeat Germany? Because the prewar British government believed in deterrence, it never thought through this kind of question, and by the time it might have been asked, there was a huge British army in France. Withdrawal would have been difficult at best. After the disaster on the Somme in 1916, many in the British government began to ask what the British should do if they were forced to accept an unsatisfactory peace, as in 1801. Part of their answer was that phase two of the war should concentrate more on the east. That is why the British had such large forces in places like the Caucasus and the Middle East when the war ended in November 1918.[7]

A century later, we are in something like the position the British occupied in 1914. We are the world's largest trading nation, and we live largely by international trade—much of which has to go by sea. We do not have a formal empire like the British, but they and we are at the core of a commercial commonwealth which is our real source of economic strength. In a crisis our trade—our lifeblood—would be guaranteed by the U.S. and allied navies, the U.S. Navy dwarfing the others. That we depend on imports means that we have vital interests in far corners of the world. It happens that relatively few Americans understand as much, or see what happens in the Far East as central to their own prosperity. Access to our trading partners there is crucial to us, just as access to overseas trading partners (and the Empire) was a life-or-death matter for the British in 1914. Like the British in 1914, we regard war as too ruinous to be worthwhile, and we often assume that other governments take a similar view. Like the British, we are not very sensitive to the possibility that other governments' views may not match ours. A long look back at 1914 may be well worth our while.

Notes

1. Michael and Eleanor Brock, eds., *H. H. Asquith: Letters to Venetia Stanley* (Oxford, UK: Oxford University Press, 1982).
2. Holger M. Herwig, *'Luxury Fleet': The German Imperial Navy 1888–1918* (London: Allen & Unwin, 1980).
3. Nicholas A. Lambert, *Planning Armageddon: British Economic Warfare and the First World War* (Cambridge, MA: Harvard University Press, 2012).
4. V. R. Berghahn, *Germany and the Approach of War in 1914*, second ed. (New York: St. Martin's Press, 1993).
5. Jack Beatty, *The Lost History of 1914: How the Great War Was Not Inevitable* (London: Bloomsbury, 2012).
6. David Fromkin, *Europe's Last Summer: Who Started the Great War in 1914?* (New York: Knopf, 2004).
7. Brock Millman, *Pessimism and British War Policy, 1916–1918* (London: Frank Cass, 2001).

8 "THE STRATEGY OF THE WORLD WAR AND THE LESSONS OF THE EFFORT OF THE UNITED STATES"

CAPT Thomas G. Frothingham, USN

Dr. Friedman's contention that World War I was a maritime war in the previous article is supported by this article that appeared in *Proceedings* in 1921—less than three years after the end of the war. Captain Frothingham's rather thorough discussion of the war as a whole is replete with evidence of the importance of naval power in the ultimate outcome.

Interestingly, he is ahead of his time in his appreciation of the importance of jointness, making it clear that the outcome is due—partially, at least—to the advantages gained by the Allies in their ability to meld various aspects of sea and land power at critical junctures of the war.

His contention that "an army and navy in being of anywhere near modern war-strength cannot be maintained by any nation on earth" is also interesting. In light of today's armed forces, one can reasonably argue that he was obviously wrong—but others might argue for his prescience when projecting into an unknown future that some contend is headed for unsustainability under current trends.

There are some cautionary warnings in his recognition of American unpreparedness early in the conflict (something that was overcome in this instance but may not always be an option in future conflicts),

> and in the need for "close relations between our Army and Navy and the American People"—a theme endorsed in the introduction to this anthology.

"THE STRATEGY OF THE WORLD WAR AND THE LESSONS OF THE EFFORT OF THE UNITED STATES"

By CAPT Thomas G. Frothingham, USN, U.S. Naval Institute *Proceedings* (May 1921): 669–83.

In the great mass of literature that has sprung up in the track of the World War, the writer has not seen any attempt to state its strategic problems in the simplest terms. It is always helpful to reduce any problem to its essential factors—and to-day the unusual condition exists that this may be done, in the case of the World War, much sooner than has been possible in studying any other war. In spite of the vastness of the struggle, never before has so much been known of a war even while it was being fought. By this is meant that there has been actual official information available, excluding the misleading writings of correspondents and critics. It is also a fact that never before have so many of the principal leaders in a war given their stories so promptly.

With these great advantages, there is no reason why we should hold off from the study of the strategy of this war on account of the large scale of its operations. The events which really counted moved with the simplicity and directness of Greek tragedy. We should realize this—and, in our amazement at the enormous volume of men and material evoked by the world's holocaust, we must not be blind to the fact that these great forces were subject to the unchanging fundamental laws of warfare. There is really nothing in the immensity of the contest that should disarm criticism, especially in view of the human fallibility shown in the conduct of this war. In fact the World War was one of blunders and unsound strategy, to a greater extent than most wars of the past.

The supposedly infallible German Superman was not long in showing ordinary human lack of understanding the initial strategic problem of Germany, and this soon neutralized the results of long years of German preparation. For Germany, at the outset, there existed an advantageous strategic situation, that was thought to be a sure promise of victory. The Teutonic Allies possessed a central and concentrated position against separated antagonists. This was the result of events in preceding years, as Russia had been shut off from Great Britain and France.

With this established condition, that their enemies would be separated, the Germans were enabled to plan to attack the Entente Allies in detail—first France, then Russia. This was the strategy of the German General Staff, and it was to be carried out by the Schlieffen plan of war "on two fronts." This plan had been determined years in advance, and it had been elaborated with rigidly fixed details, to the exclusion of all other solutions. Its strategy obsessed the Germans. It was the product of the hierarchy of Clausewitz, Moltke, and Schlieffen—and that it would fail was thought impossible.

Of this German plan of war, it should be stated at once that it violated a fundamental of warfare, in that it was essentially a military plan and neglected to make full use of the navy arm. Admiral Tirpitz unqualifiedly says that the navy's plan of operations "had not been arranged in advance with the army." For the Germans to allow themselves to be absorbed in this military plan was an error.

It is true preponderance in sea power rested with the Entente Allies, but Germany had made great efforts to develop a navy. The German fleet possessed the advantages of the double base at Kiel and the outpost of Heligoland. The Germans had also made great advances in the use of mines and submarines, the weapons of a weaker fleet. In addition, there was another argument for naval activity in cooperation with their armies. Outside of the recognized value of additional threats as diversions at the time of an intended offensive, there was a state of mind in Great Britain which would have favored the success of German naval threats. It is known that belief in a German plan to invade England was prevalent among high British officials, and this would have helped a German naval diversion. In this respect, at the outset, the German High Command failed to use all the strategic means possessed by the Central Powers. As a matter of fact, the German naval forces merely made scattered forays of no importance, and at the outbreak of war German strategy must be studied as restricted to the Schlieffen military plan—to overwhelm France by "forcing a speedy decision," while Russia was to be contained.

Yet it must be admitted that, if the German purely military solution had proved to be correct, there would have been grave danger of a defeat that would have paralyzed France and lost the war before the sea power of the Entente Allies could have exerted its effect.

To appreciate this danger it is necessary to realize that, at the declaration of war, as a result of the preparations of Germany, a military situation actually existed which enabled the Central Powers to call into being a superior military force at the outset. On the other hand, time was required by the Entente Allies to produce an equal force and to make use of sea power.

Consequently the logical strategic aim for Germany should have been to impose this superior military force at once with destructive effect upon the Allied armies. This long-prepared Teutonic military superiority would only be able to gain its necessary quick victory, if brought into immediate crushing contact with the less prepared armies of the Allies. It is an axiom that the stronger force should seek contact at once with a weaker enemy, and not allow the weaker enemy time to gather strength. Immediate contact with the French armies should have been the essential of German strategy.

The Schlieffen plan of war had been devised before 1906, in the days when it was held that against France "a frontal attack offered no hopes of success" on account of the high value assigned to the French frontier fortresses as strong points in the positions of armies. To avoid a frontal attack Schlieffen planned the encircling sweep through Belgium. In 1914 the Germans actually possessed new artillery which neutralized fortresses as strong points, but the German strategy had become so tied to this fixed plan that there was no thought of change, and it was carried out in every detail.

This elaborate encircling movement through Belgium, to which the German General Staff had thus committed the powerful German armies, failed to produce the essential result of imposing these German armies in destructive contact upon the Allied armies, until after the Allies had gathered sufficient force to fight an equal battle—and, in the words of Falkenhayn, "the intention of forcing a speedy decision, which had hitherto been the foundation of the German plan of campaign, had failed."

It should also be stated that, although the German armies were mobilized with wonderful efficiency, the German Headquarters made a poor showing in handling these armies in their scheduled operations. Owing to the lack of army group control the two encircling armies of Kluck and Bülow were at odds a great deal of the time.

On the Eastern front the German General Staff had overconfidently assumed that Russia would be slow in mobilizing and would be contained without much difficulty. Instead of this, East Prussia and Galicia had been invaded, and, although Hindenburg and Ludendorff had decisively defeated the invasion of East Prussia, the Eastern front had already become a drain upon the resources of the Central Powers.

Consequently the middle of September, 1914, saw the failure of German strategy to win decisive results with the great forces that had been prepared through so many years. The Moltke regime was ended, though this fact was kept secret to prevent "further ostensible proof of the completeness of the victory obtained on the Marne." Then and there the decision was forced that the World War was not to be a quick overwhelming victory won by Germany's long-prepared military strength. Not only had the perfected strategy of the Schlieffen school failed to win the victory which had been thought certain, but also great harm had been done to the prospects of Germany by the moral effect of the invasion of Belgium. Moral forces are of actual strategic value in war—and there is no question of the fact that the violation of Belgium arrayed strong moral forces against Germany.

By this defeat of the German war plan of 1914 a complete change had been brought about in the strategic situation. The Central Powers had lost the offensive. Not only had their armies been brought to a stand-still, but the Teutons were ringed about by their enemies, and from this time on they were destined to feel the relentless pressure of sea power in the hands of the Allies. At this stage the Central Powers were practically besieged, being even shut off from their new ally, Turkey. This situation per force created its new strategic objectives. That of the Allies was to constrict and press the siege. That of the Central Powers was to break through and raise the siege.

These objectives could hardly fail to be the visible motives of the strategy of 1915. The advantage lay with the Allies, and on the surface their plans were promising—the attack upon the Dardanelles, followed by prepared offensives in the West and in the East, with the entrance of Italy against Austria. It was in their estimate of the means necessary to carry out their strategy that the Allies failed utterly. In the Dardanelles attack ships' guns were expected to equal the effect of the siege howitzers, and the other offensives of the Allies so entirely lacked any conception of the forces necessary, that they became mere ineffectual nibbles that did not even divert the Teutons from their objectives.

On the other hand, it must be admitted that, in contrast with 1914, the new regime in the German General Staff showed an adequate conception of its necessary strategy. Yielding to the Hindenburg-Ludendorff influence, it was planned merely to hold on the West but to break through by main strength of artillery in the East, a great advance over the artificial strategy of 1914. In 1915 the German strategy was complete both in preparation and in the tactics employed. The break-through was accomplished with losses for the Russians that crippled their armies beyond repair, and it also resulted in the accession of the Bulgarians and the overthrow of Serbia. These victories were won without a possibility of a diversion, in consequence of the feeble preparations of the Entente Allies. The siege of the Central Powers was raised, and the Mittel Europa tract was won by the Teutons.

Only on the seas had the Entente Allies won results in 1915. The German fleet had remained a menace that necessitated a great force to contain it; the Germans controlled the Baltic and the Dardanelles—and in 1915 were making tentative efforts to develop a submarine offensive. But in the main the Allies controlled the seas, reaped the advantages of free use of the water-ways, and excluded the Central Powers from use of the seas—thus keeping up a pressure upon the Teutonic Allies that increased as time went on.

At the beginning of 1916 there was a great strategic opportunity for the Central Powers. Russia had been crippled to such an extent that evidently it was out of the question for the Russians to undertake an offensive in the early part of the year. The measure of the other Allies had been taken in their weak efforts of 1915, and consequently it was also certain that no offensive was to be feared by the Germans early in 1916 in other theatres. The Central Allies were thus enabled to take the offensive and to concentrate troops at chosen points of attack, without running the risk of counter strokes.

It was decided that the Germans would attack at Verdun and the Austrians would undertake an offensive against the Italians. These offensives were both sound operations in every strategic sense to undertake at this time, with conditions in favor of their success as stated—and with decisive results sure to follow, if carried out by proper means.

Yet with all this advantage for the Teutons in military conditions established at the beginning of 1916, the Central Powers failed to adopt sufficient means to carry out their strategy.

The German assault upon the Verdun sector of the Western front was planned to duplicate the successful break-through by concentration of heavy artillery against the Russians in 1915. As Falkenhayn states, it aimed to impose the German force at a place where the French must stand and fight. The concentration of artillery and troops was successfully made, with the element of surprise attained—and the immediate success of the initial attack showed the soundness of the plan, if it had been made upon a broader scale. As it was, in their confidence in the German artillery tactics for breaking through, the Germans carried out their operation on so narrow a sector that their offensive was smothered. The small margin by which it failed, even with this defect, is comment upon the danger if anything approaching the German tactics of 1918 had been adopted.

The Austrian offensive of 1916 against the Italians had also won a measure of success, but the fatal error had been made of over-confidently weakening the southeastern front opposing the Russian armies. The Russians were thus enabled to take the offensive in June, and win sufficient success against the weakened Austrian lines to compel the Germans and Austrians to send reinforcements to that front. This necessary diversion of troops made it impossible to go on with the offensive against the Italians.

The Central Powers had thus lost the offensive on both fronts, and at the middle of the year 1916 there was a complete change in the military situation. As a result, in the last half of 1916, the Entente Allies were enabled to undertake three ambitious offensives, the Battle of the Somme, the Russian attacks in the southeast, and the Italian Gorizia offensive. From all of these great things were expected, but none fulfilled the hopes of the new optimism then prevalent among the Allies.

Of these offensives the Battle of the Somme was made possible because the British armies raised by conscription were at last ready for action on the first of July. On that day began the series of joint operations by the British and French, with the British taking the major part, which lasted into November. The Battle of the Somme was a determined effort to dislocate the German armies, but tactics had not been devised that would bring to bear forces strong enough to win that result, and the operation degenerated into the so-called tactics of "limited objectives," with heavy losses for the British.

The power of the Russian offensive had been much overestimated, and the Russian armies were checked all along the line by early fall. The same fate befell the Italians in their offensive. After capturing Gorizia without much trouble, the Italian armies were again brought to a standstill in the difficult mountainous country.

Rumania had also made her ill-timed entry, having delayed until after the Russians had been rendered incapable of rendering assistance. Consequently the new Hindenburg-Ludendorff regime, which had been given control of the Austro-German armies, had no difficulty in overrunning Rumania, while at the same time the Somme offensive was worn down, and the Italians and Russians checked from any operations that might have created a diversion—an example of the strategic value of central position and united control.

On the sea in 1916 the Entente Allies were exerting increasing pressure upon the Central Powers. The great indecisive naval action of Jutland had brought no change in the actual situation. Germany had attempted an illegal campaign with her submarines, but this had been dropped upon the demand of the United States. At the end of 1916, urged on by those who held that unrestricted submarine warfare would win the war, and convinced of the truth of their calculations, the German High Command adopted the fateful plan of war of 1917—to hold their conquests with their armies, but to make their offensive an illegal campaign of unrestricted submarine warfare.

From this date Germany made herself an outlaw among nations—and again the German High Command had deliberately chosen strategy that aligned strong moral forces against the nation. The German leaders had founded their strategy of 1917 upon the calculation that they would win before outraged America could change the result. They thought they could win by these foul means—and the result of their decision was upon their own heads.

The campaign of unrestricted submarine warfare proved successful to a degree that upset existing conditions. It came very near destroying the Allied control of the seas and winning the war, but it also accomplished the most harmful strategic result of all against Germany, as it brought the United States into the war—a factor that turned the balance to defeat for the Central Powers.

On the other hand, at the beginning of 1917, the Allies' conception of their strategy was utterly at variance with the imposed existing conditions. At last the Entente had advanced to the stage of a plan for concerted military operations on all fronts. Yet the actual situation was that Russia had been so broken, in bearing the main burden of the war for two years, that the nation was on the point of revolution and was no longer of any value as a military factor. As a result, in the spring of 1917, the strategy of the Entente Allies, with Russia paralyzed by Revolution, had dwindled into attacks upon the Western front, made more difficult by the rectification of the German lines by the Hindenburg-Ludendorff control.

Fortunately the completeness of the military collapse of worn out Russia had not been appreciated by either side. For this reason, the gradual development of Russian helplessness was less of a shock to the morale of the Allies, and the Central Powers were slow in realizing that they might safely withdraw large numbers of troops from the Russian front. In fact the last strategic service of the Russian armies was thus rendered by the empty threat of their former power, and the delay made the concentration against the Italians so late in the year that it helped to save the Italian armies from utter destruction in the disaster of Caporetto. But the year 1917 ended with the Italian armies so shattered that to repair them had become a drain upon Great Britain and France, at a time when the British and French armies had been woefully depleted by the losses of the unsuccessful battles of 1917 on the Western front.

It was true that the Central Powers had failed to win their expected decision through unrestricted submarine warfare, but the beginning of 1918 found them enabled to concentrate the full German strength upon the Western front, without any danger of a diversion elsewhere. This ability to move troops from the East gave the Germans an actual superiority in numbers, as the British and French resources in man-power had been drained in the costly and unsuccessful battles of 1917, to such an extent that it had become a hard task to fill the ranks of the British and French armies. There was no hope of an increase to offset the German reinforcements from the East.

Possessing this assured superiority the Germans were able to plan their offensive of 1918 without any danger of counter attacks. Ludendorff had become the controlling power in the German General Staff. His strategy was a return to the direct methods of concentration of forces against a chosen point of attack, and new tactics had been devised by which many divisions were grouped against the chosen point, insuring successive streams of troops which infiltrated the enemy positions and dislocated the defenders.

These new tactics were surprisingly effective against the Allies, and at the beginning of July, 1918, this formidable German offensive, in a series of overwhelming attacks, had so smashed and dislocated the Allied armies, even after they had at last been united under the command of Foch, that it is difficult to see how the situation could have been saved except by a strong reinforcement for the Allies—and this could only be furnished by the American troops.

To define this critical military situation explicitly, it is only necessary to quote the following statement of the Versailles Conference, June 12, 1918:

"General Foch has presented to us a statement of the utmost gravity . . . as there is no possibility of the British and French increasing the numbers of their divisions . . . there is a great danger of the war being lost unless the numerical inferiority of the Allies can be remedied as rapidly as possible by the advent of American troops . . . We are satisfied that General Foch . . . is not overestimating the needs of the case . . ."

D. Lloyd George.
Clemenceau.
Orlando.

The United States Army was able to provide the necessary reinforcement that turned the balance, with the result that the offensive always remained against the Germans, and they lost the war.

There is no longer any question of the fact that the German Headquarters made their calculation that it was utterly out of the question for the United States to exert any physical force upon the war. The German leaders had on occasions yielded to keep us out of the war, to avoid having our resources at the service of the Allies, but the Germans applied their own formulas to our nation, and, following these, it was held a military impossibility for an adequate American army to appear upon the fighting front. It must also be said that this was the prevailing opinion among European military experts of all countries—and from the European point of view a military impossibility was accomplished when our troops performed their part in the war.

Our strategic problem was an operation against a contained enemy—with the great advantage for us of freedom from danger of being attacked. But it was complicated by the condition that transportation overseas, which would normally have been provided by Allied shipping, had been impaired by the submarines to so great an extent that we were compelled to provide a large share of the transportation ourselves. The submarine menace, and its diversion of Allied naval forces, also made it imperative for us to provide a great proportion of the necessary naval protection. There was the added urgent necessity of haste—or the war would be lost.

This crisis demanded an effort on the part of the United States that would comprise: raising and training an army; transporting a great part of that army overseas; providing supplies and transporting them overseas; giving naval protection; providing terminals and bases overseas to receive and handle the troops and supplies. All this must be done in haste, and at the outset on the large scale set by the unprecedented demands of the World War. There was no time for the gradual development of forces, as in the case of other nations.

No nation in history ever faced such a task, and all this was accomplished by the surge of our people, united in belief in our unselfish duty in the war—a force moral as well as physical that brought about cleavage between the German Government and the German people, which became a strong factor in breaking down the German militaristic structure. Our moral force sowed the seeds of German revolt against the German Government—and America's unexpected physical strength for war turned German victory into German defeat.

In tracing the course of the war, the failure is self-evident of the most perfected military machine in all history—and the continued inability of the Allies to progress beyond piece meal methods is equally apparent. The wonder of the war has been the fact that the peaceful United States proved to be the one nation that coordinated the functions of its military, naval, and industrial forces, to accomplish its full strategic objective, in the time set by a crisis and on the enormous scale demanded by the World War.

To study the causes that brought about this result will become the most interesting thing in connection with the war. Our effort will be recognized as one of the great uprisings, which have shown the world that human forces united by some powerful fusing impulse are stronger than artificial military conditions. To find a comparison, with the exception of our Civil War, it will be necessary to go to the great movements of the northern races which overran Europe. France, after the Revolution, has always been considered unique as an example of a united uprising of humanity finding in Napoleon an ideal leader. Yet, with all the years of enthusiasm for the Emperor, it was only the military and industrial forces that reached full strength—Napoleon was never able to vitalize the naval arm.

It should be bluntly stated that, in every military sense, we were unprepared—and this retarded everything at the start. For a time it looked as if European prophesies as to our helplessness in war would prove true. Then, from delays and confusion emerged the miracle, the army and navy forces of the United States. It is true that all kinds of mistakes were made, but behind our operation was a strong impelling force that had not been measured since the Civil War.

As has been said, the Civil War is the only basis for comparison. In that war our nation had shown that Americans, when aroused by an appeal, instinctively developed strategy, tactics, and weapons far in advance of their time. Students of the Civil War believed that the qualities shown in that epoch-making war were still innate in our people, but European experts had never appreciated the lessons of 1865 until the World War had confirmed them—and there was even doubt in America as to whether the same fibre remained in our nation augmented by immigration.

But, at the great summons, it was shown that the same spirit was vital in America. We had even advanced, as a result of the American habit of mind in thinking in terms of great masses in all our industries. This made it instinctive for Americans to solve our war problems by means of the same methods, of assembling the great plants first and then their products, in men and in material, on a large scale. These American methods insured the success of our effort on land and sea.

The striking attribute of the Civil War soldier or sailor, the game-playing ability to handle himself and his utensils, was again present. With our freedom from the dulling effect of class distinctions, there was intelligence in all ranks, and a better coordination, with, above all things, the adaptability of the Americans to receive instruction from infusion of trained personnel.

In fact one great outstanding lesson of the World War was the demonstrated fact that it was possible to vitalize a new American personnel in the Army and Navy by the infusion of a comparatively small number of trained men, and in an incredibly short time the whole personnel would be "leavened," as Grant expressed it. It is no exaggeration to state that, given a selective draft of the American intelligent and adaptable personnel, it was proved possible in the U.S. Army and in the U.S. Navy for every nucleus from a trained force to produce a unit of practically equal value.

The effect of this goes far beyond any formerly accepted basis for estimating our defensive strength. The conclusion is logically suggested that the United States has a resource which, rightly used, would enable our country to be the first to produce forces on the vast scale demanded by warfare of today. This is speaking in terms of the present, when the establishments of Europe have been scattered. At once a greatly increased value is shown for our Army and Navy in this enlarged function of infusion of skilled elements into a new personnel.

Taking the only broad view of this developed function of the U.S. Army and U.S. Navy, it is evident that there are two fixed conditions in the problem of the United States. On the one hand, in view of the lessons of the war, an army and navy in being of anywhere near modern war-strength cannot be maintained by any nation on earth. On the other hand, the United States has shown that it has the quickest turnover of man-power in the world, especially adapted to receive training through contact with an infusion of skilled officers and men. This at once gives a double value to our Army and Navy in being, not only as trained forces in themselves, but as leaven for the great mass of citizen recruits. In no other nation is it possible for the permanent establishment to have this second value. Consequently the United States has the best argument for maintaining its Army and Navy.

This new value should be recognized as an asset for our nation, of which the test has been made in war. Especially should we realize its existence at the present time, when we are at the stage of fault finding and recrimination, with every mistake being magnified and every personal grudge expressed. Added to all this, is the natural reaction and loss of interest in army and navy affairs. This is only a temporary apathy, but there is danger of the Army and Navy again becoming separated from our people.

Before the World War the services and the people were strangers. Our citizens did not understand the Army and the Navy, and both services underestimated the capacity of our people for warfare. The World War has brought every family in the country in touch either with the Army or with the Navy, and it has been proved that in this close relationship lies our strength. It is important that this touch should be maintained, not as a task imposed upon our citizens, but with the recognition of mutual interest and ultimate service together. This is the doctrine taught by the war.

It is consequently most important that every means should be employed to maintain close relations between our Army and Navy and the American people. A working basis has been established for this in the new Army policies of localized units and vocational training. By these provisions the soldier is identified with some definite community, and the young man in the army is being made valuable to his community in civil life, even while he is drawing army pay. In the Navy good results in the same line are to be expected from the new service for Reserves and from the close relationship with the enlarged Merchant Marine.

The R.O.T.C. courses are very practical means of keeping our young men in touch with the services, and every young man who takes such a course not only has the benefit of the training himself, but he takes a knowledge of the service home with him, and thus becomes a link between the services and the community. For this reason every form of voluntary study should be encouraged—and camps and practice cruises provided.

The National Guard is another element of great value, and as great a number as possible of Reserves should be maintained. Both factors are important links between our people and the services, and the friendliest relations should be established for mutual understanding.

With the recognition that the defense of our country means welding our people, our Army, and our Navy into one whole, the United States has the highest and truest reasons of all the nations on earth for maintaining an army and a navy. We accomplished our task in the World War because our Army, our Navy, and our industries all worked together, impelled by the great fusing force of our national uprising. We should never again look upon them as separate factors—and we never should forget the source of our strength. The lesson we have learned should be the foundation of our doctrine, never to be obscured by other interests.

9 "GETTING SEA CONTROL RIGHT"

Milan Vego

A renowned strategic thinker, Dr. Vego warns that although sea control is one of the most important objectives in a war at sea, modern Navy planners often fail to fully understand the true meaning of this important concept. He differentiates sea control from sea denial, contending that the two are distinctly different concepts, requiring different methods of accomplishment. Arguing the continued relevance of past naval thinkers like Mahan, Corbett, and Castex, he advocates relevant education for today's officers that will embrace classical theories and ensure that such concepts as sea control are understood and properly used.

"GETTING SEA CONTROL RIGHT"
By Milan Vego, U.S. Naval Institute *Proceedings* (November 2013): 64–69.

The U.S. Navy's failure to understand sea control and sea denial is more than mere semantics—it could negatively impact the development of future operational concepts.

Although it is one of the most critical objectives in a war at sea, the U.S. Navy has difficulty properly understanding the true meaning of sea control and that of its counterpart, sea denial. Often sea control is confused with naval capabilities, and for the most part the service's current doctrine and posture statements do more to obfuscate than clarify the purpose, attributes, and primary methods for obtaining, maintaining, and exercising sea control. Additionally, the Navy does not seriously consider sea denial as a possibility in a case of war with a strong opponent at sea.

In the past, the principal objective of a stronger fleet in a war was to obtain and maintain command of the sea in a given maritime theater. Such command aimed to ensure the free use of sea communications and to fully deny its use by the enemy. Narrowly defined, command of the sea was understood to be nothing more than command of sea routes.[1]

Following the realization that mines, torpedoes, submarines, and aircraft made it difficult—even for the stronger navy—to obtain absolute and permanent command of the sea for any extended time over a large part of the theater, the term sea control came gradually into use. Broadly defined, sea control refers to the ability to use a given part of a body of water and its associated air space for military and nonmilitary purposes in time of open hostilities. It more accurately conveys the reality that in a war at sea between two strong opponents, it is not possible—except in the most limited sense—to completely control the seas for one's use or to completely deny an opponent's use.

While sea control is an offensive objective, sea denial is invariably a defensive objective at the strategic level and is the principal objective of the weaker side. It aims to deny in part or full an adversary's use of the sea for military and commercial purposes. The weaker side, however, may transition to the offensive at the operational and tactical levels. In some cases, a stronger side could temporarily be forced to be on the defensive strategically, either because of the large losses suffered in the initial phase of a war or its inability to be on offensive in a given maritime theater or in two wide separate theaters.

Current Doctrine

The 2010 *Naval Doctrine Publication 1: Naval Warfare* (NDP-1 2010) notes that the purpose of sea control is to allow U.S. naval forces "to close within striking distance to remove landward threats to access, which in turn enhances freedom of action at sea" and enables "the projection of forces ashore."[2] This thinking, however, puts the cart before the horse. Enemy land-based aircraft, air defenses, and coastal defenses must be destroyed or neutralized before sea control is actually obtained. The 2010 *Naval Operations Concept: Implementing the Maritime Strategy* (NOC 10) contends that "our [U.S.] ability to establish local sea control is fundamental to exploiting the maritime domain as maneuver space, protecting critical sea communications, and projecting and sustaining combat power overseas."[3]

In addition to denying the enemy military and commercial use of the sea, the true purposes of sea control are to provide support to friendly ground forces operating on the coast, pose the threat of attack—and actually attack—the enemy's shore, and ensure uninterrupted flow of friendly military and commercial shipping. By possessing sea control, the stronger side can have a major, and often decisive, influence on the course and outcome of war on land. Sea control also can lead to drastic shifts in political and military alignments in the maritime theater.

Misconceptions and Their Problems

In the U.S. Navy, sea control is often considered essentially the same as sea denial. While it is true that possessing sea control means at the same time denying that control to one's opponent, there is a fundamental difference between these two concepts. Theoretically, the act of obtaining or gaining something is a positive objective, while denying or preventing it is a negative objective. From this stems different operational realities, as the methods used for obtaining sea control are considerably different from those used in sea denial.

Viewing sea control primarily as a capability and not as an objective to be accomplished, the Navy, in its 2010 capstone doctrinal document, NDP-1, includes sea control along with forward presence, deterrence, power projection, maritime security, and humanitarian assistance/disaster relief (HA/DR) among its core capabilities.[4] The doctrine explains that power projection "in and from the maritime domain includes a broad spectrum of offensive military operations (aimed) to destroy enemy forces or logistics support or to prevent enemy forces from approaching within enemy weapons range of friendly forces."[5] Specifically, power projection pertains to "the ability of a nation to apply all or some of its elements of national power—political, economic, informational or military—to rapidly and effectively deploy and sustain forces in and from multiple dispersed locations" in responding to a crisis, contributing to deterrence, and enhancing regional stability. It consists of two components: sea power and air power.[6] Forward presence of naval forces is an integral part of power projection, yet it does not follow that power projection itself consists of actions aimed at obtaining or denying sea control.

The more serious problem, though, is that while the Navy firmly believes sea control exists in peacetime by virtue of its combat potential, reality tells a different story. If the Navy possesses and exercises sea control in peacetime, how was it possible for U.S. naval vessels to be harassed with impunity by five high-speed Iranian boats in the Strait of Hormuz in January 2012? Moreover, during the Syrian crisis, in September 2013 a Russian helicopter carrier and five other warships operated in the proximity of five U.S. destroyers and an amphibious ship. At the same time, China sent an amphibious dock landing ship and other vessels to observe U.S. and Russian actions off Syria's coast.[7] In peacetime, sea control is not a prerequisite for the service's forward presence, nor it is required for providing HA/DR. It may, however, be required in some operations, such as supporting counterinsurgency or conducting counter-piracy.

In reality, during peacetime, no navy possesses sea control, but instead only exerts a certain degree of *naval influence*. Both naval combat potential and naval influence are relative, and no clear relationship exists between the capability to act and the probability of exercising influence. Naval influence is more subtle and more ambiguous than naval combat potential. Forward presence is only one of several aspects of naval influence. For example, a weaker navy operating from a more favorable geographic position might have greater naval influence than the U.S. Navy, which must project power over distances of several thousands of miles.

Naval influence is the sum of one's naval combat potential and non-naval capabilities, such as air forces and land-based ballistic or cruise missiles as well as the quality of leadership and personnel. The Chinese navy, by augmenting the combat potential of its naval forces by air forces and land-based medium-/short-range ballistic missiles, might have a greater naval influence in the western Pacific than the U.S. Navy. Naval influence, though, encompasses more than just hardware and includes many difficult or impossible to measure elements.

The perceptions of one's naval combat potential and willingness to use it in time of crisis often differ among the friends, potential enemies, and neutral parties in a given maritime theater. Naval influence is generally greater if the stronger side at sea has a history of following through on its statements with actions rather than merely issuing empty threats. If, for instance, China were to establish sovereignty on some, or all, of the disputed lands in the South China or East China seas it may benefit from greater naval influence in territorial disputes than the United States, which may be unwilling to assume a more forceful stance to protect the interests of its friends and allies.

Degrees of Control

Too broad and imprecise, the term sea control fails to appropriately convey the reality that in a war between two strong opponents sea control is relative. One of Sir Julian S. Corbett's greatest contributions to naval theory is his emphasis that command of the sea "may exist in various degrees" but that it "can never in practice be absolute."[8] Likewise, French strategic theorist Vice Admiral Raoul Castex argued that command of the sea is "not absolute but relative, incomplete and imperfect."[9] Sea control can be general and/or local. General sea control pertains to a state in which the stronger side exercises a rather loose and incomplete control of a large part of a maritime area without significant challenge from its opponent.[10] Local sea control exists when one side has high degree of control in a relatively small part of a maritime theater, such as the amphibious objective area, that is considered critical for accomplishing an operational objective.

Sea control encompasses control of the sea surface, subsurface, and associated air, with overall control depending on the degree of control over each of these three dimensions.[11] Control of the air greatly affects control of the surface, and in many ways the control of subsurface. In the same way, control of subsurface affects control of surface, and to some extent control of the air. For example, nuclear-powered attack submarines could be used to suppress an enemy's ground-based air defenses. In the information era, control of cyberspace, which is neither absolute nor permanent and exists only in degrees, plays a part, too.

The struggle for sea control, or denying control, cannot be successful then without full unity of command. This means that a joint force maritime component commander, supported by a joint force air component commander, must be solely responsible for planning and executing major naval/joint operations and other tactical actions in this effort.

Sea control can be permanent or temporary, limited or absolute. Permanent sea control exists when the stronger side completely dominates a given maritime theater, either because the other side does not have any means to deny that control or because its fleet has been destroyed. Corbett wrote that permanent sea control does not mean that the opponent can do nothing, but rather that it cannot interfere with shipping or amphibious landings in such a way as to seriously affect the course of the war.[12] In practice, however, it is more common that the weaker side still has some means at its disposal to challenge the stronger side's control.

Temporary sea control exists when one side possesses a high degree of control over surface, subsurface, and air for a limited time. When possessing limited sea control one side has a high degree of freedom to act on the sea while the other side operates with high risk. It is inherently transitory and unstable.[13] Absolute sea control means that a state's naval forces can operate without major opposition, and their adversary cannot operate at all.

Winning the Struggle

The fight encompasses three distinct but closely related and overlapping phases: obtaining, maintaining, and exercising. In operational terms, the strategic objective is accomplished by obtaining sea control, which is followed by one's efforts to consolidate strategic or operational success to maintain control through destruction of the remaining enemy's forces. Exercising sea control entails exploiting these strategic or operational successes. There are no clear lines separating these phases: Some actions are predominantly carried out during the exercising phase and are conducted as soon as the sufficient degree of sea control is obtained.

Interestingly, there are no specifics outlined in the U.S. Navy's doctrine as to how to obtain and maintain sea control, nor do they use the terms *exercising* or *exploiting*, but instead focus almost exclusively on *power projection*. For example, the NOC 10 explains that U.S. naval forces would "achieve sea control by neutralizing or destroying threats in the maritime, space, and cyberspace domains that constrain our freedom to maneuver, conduct follow-on missions, or restore maritime security."[14] Generally, the main methods for obtaining sea control are destruction and/or containment (or neutralization) of the enemy's naval- and/or land-based air forces; weakening the enemy's naval forces over time; seizing control of choke points; and capturing the enemy's naval- or air-basing areas. Exercising sea control includes both the threat and execution of amphibious landings on the unopposed or opposed shore, destroying the enemy's coastal area and facilities/installations, conducting commercial blockade, and providing support to friendly ground forces in their offensive (or sometimes defensive) operations on the coast.

The principal methods of combat employment of naval forces in obtaining, maintaining (or denying), and exercising sea control are tactical actions and major naval/joint operations. Tactical actions, such as attacks, strikes, raids, engagements, and battles, aim to accomplish minor or major objectives. In contrast, a major naval operation consists of a series of related tactical actions sequenced and synchronized to accomplish an operational, and in some cases a partial strategic objective. They are planned and conducted by a single commander in accordance with a common operational idea. In the littorals, major naval operations will be conducted in cooperation with sister services and the services of other nations. Each major naval/joint operation would strive to have a decisive impact on the course and outcome of the struggle for sea control. In addition, it would be necessary to conduct numerous tactical actions within a given operational framework.

One of the enduring and most serious problems is the failure to recognize major naval/joint operations as the principal method for U.S. naval forces to accomplish operational objectives at sea in wartime. For example, NDP-1 2010 and NOC 10 refer to a sea control operation as "the employment of naval forces, supported by land, air, and other forces as appropriate, in order to achieve military objectives in vital sea area."[15] However, this definition does not include either tactical methods or the scale of the objective to be accomplished. It is also too broad because it encompasses multiple objectives, which would require a major naval/joint operation. Both documents also specify that a sea control operation would include "destruction of enemy naval forces, suppression of enemy sea commerce, protection of vital sea lanes, and establishment of local military superiority in area of naval operations."[16] "Suppression of enemy sea commerce" is not a method for obtaining sea control, but an objective to weaken the enemy's military-economic potential.

The NOC 10 explains that U.S. naval forces conduct sea control operations in environments ranging from uncertain to openly hostile, where they often contend with adversarial tactics such as opposed transit, anti-access, and area denial.[17] The authors of the document obviously consider sea control operations as the method to be used in peacetime as well and assert that U.S. "naval forces will for the foreseeable future conduct sea control operations to enforce freedom of navigation, sustain unhindered global maritime commerce, prevent or limit the spread of conflict, and prevail in war."[18] In what seems to be purely tactics, NDP-1 2010 specified that sea control operations "involve locating, identifying and dealing with a variety of contacts."[19]

The U.S. Navy's current understanding of sea control and sea denial needs to be brought in line with widely accepted views, and it must be recognized that these are two distinct concepts to be accomplished through different methods. As the service reduces the size of its battle force, it must understand that it may be forced, at least temporarily, to be strategically on defensive, that is, to conduct sea denial. At the heart of any sound doctrine at the operational level of a war at sea are operational concepts. Although U.S. joint doctrine recognizes major operations as the primary method of combat employment to accomplish operational objectives, the U.S. Navy does not and focuses on so-called strike warfare, or tactics. Such a disconnect will likely negatively affect the development of needed operational concepts. Despite the passage of time, the views of naval classical theoreticians, notably Mahan, Corbett, and Castex remain highly relevant in our information age, and the service should pay far more attention to the education of its officers in the study of naval and military history, and naval theory in particular. Using proper operational terms and fully understanding their true meaning is not a matter of semantics but a prerequisite for an effective communication within a service and among services.

Notes

1. Otto Groos, *Seekriegslehren Im Lichte Des Weltkrieges. Ein Buch fuer den Seemann, Soldaten und Staatsmann* (Munich: E. S. Mittler & Sohn, 1929), 43.
2. *Naval Doctrine Publication 1: Naval Warfare* (Washington, DC: Department of the Navy, March 2010), 27.
3. *Naval Operations Concept 2010: Implementing the Maritime Strategy* (Washington, DC: Department of the Navy, 2010), 53.
4. *Naval Doctrine Publication 1: Naval Warfare*, 25.
5. Ibid., 29.
6. *Naval Operations Concept 2010: Implementing the Maritime Strategy*, 60.
7. "Russian Carrier Scheduled to Sail to Syrian Coast," *Military.com,* September 4, 2013, www.military.com/daily-news/2013/09/04russian-carrier-scheduled-to-sail-syrian-coast; Mark Adomanis, "Russian Ships off Syria Will Likely Do Little," *USNI News,* September 9, 2013, news.usni.org/2013/09/09russian-ships-will-likely-little.
8. Julian S. Corbett, *Some Principles of Maritime Strategy* (London: Longman, Green and Co., 1918), 89–90.
9. Raoul Castex, *Strategic Theories*. Selections translated and edited with an introduction by Eugenia C. Kiesling (Annapolis, MD: Naval Institute Press, 1994), 53
10. Guenther Poeschel, "Ueber die Seeherrschaft (II)," *Militaerwesen* (East Berlin) (June 6, 1982): 72.
11. Guenther Poeschel, "Ueber die Seeherrschaft (I)," *Militaerwesen* (East Berlin) (May 5, 1982): 42.
12. Corbett, 91.
13. Guenther Poeschel, "Ueber die Seeherrschaft (II)," 71–72.
14. *Naval Operations Concept 2010: Implementing the Maritime Strategy*, 57.
15. *Naval Doctrine Publication 1: Naval Warfare,* 27–28; *Naval Operations Concept 2010: Implementing the Maritime Strategy,* 51–52
16. *Naval Doctrine Publication 1: Naval Warfare*, 28; *Naval Operations Concept 2010: Implementing the Maritime Strategy,* 51–52; Joint Publication 1–02: *Department of Defense Dictionary of Military and Associated Terms*, November 8, 2010 (As Amended Through March 15, 2013), 252–53.
17. *Naval Operations Concept 2010: Implementing the Maritime Strategy*, 53–54.
18. Ibid., 52.
19. *Naval Doctrine Publication 1: Naval Warfare*, 28.

10 "CEDE NO WATER: STRATEGY, LITTORALS, AND FLOTILLAS"

CAPT Robert C. Rubel, USN (Ret.)

Few contemporary students of naval strategy have not heard of "Barney" Rubel. His relevance to the field is both legend and real. In this article, he delves into the contemporary issue of littoral combat by urging a return to the flotilla concept, arguing that this seemingly passé approach has relevance in the littorals today. He confronts "some cultural and emotional Rubicons for the Navy to cross in order to achieve effective flotilla operations" and builds a compelling case for this seemingly anachronistic approach, pointing out that it contributed to Union victory in the Civil War: "While the ships have changed, the basic principle of maneuverable vessels for tight environs remains the same."

"CEDE NO WATER: STRATEGY, LITTORALS, AND FLOTILLAS"

By CAPT Robert C. Rubel, USN (Ret.), U.S. Naval Institute *Proceedings* (September 2013): 40–45.

For future naval activity that involves closer proximity to potential adversaries' coastlines—and thus a requirement for more maneuverable ships—the Navy should consider flotilla operations.

Debate has been raging over the merits and drawbacks of the littoral combat ship (LCS). Critics regard it as too big, too expensive, too vulnerable, and not sufficiently capable overall. Even its defenders argue that it was never designed to operate in highly threatened waters. Some advocate the development of much smaller combatants—small craft, almost—that would operate in the most dangerous areas.[1] The problem with the current discussion is that it mostly revolves around ship characteristics, with little or no thought given to the strategic issues involved, or is based on a bunch of unexamined assumptions about why U.S. Navy ships would be in such waters in the first place. This is an attempt to bring strategic logic to bear on the matter of U.S. naval operations in the littorals and confined seas in the hope that it will provide insights that will inform force design, including ship and weapon characteristics, organization, and personnel management.

Why American ships would be in a contested littoral is a matter of naval strategy. Broadly speaking, the U.S. naval strategy is to disperse the Fleet to patrol potential hot spots along the Eurasian littoral to deter trouble and be readily available as first responders in case it arises. Since Operation Desert Storm, such patrols have been carried out in the absence of any substantial opposition or threat at sea.

Now, as so many writers have pointed out, the growing navies and assertive actions and rhetoric of China and Iran force the U.S. Navy to take account of new conditions, especially in the littorals and confined waters adjacent to those nations. It would be easy enough to simply assume that U.S. maritime strategy will remain constant and regard the issue as a matter requiring only technical adjustment; i.e., either better defenses for current classes of ships or new operational concepts such as Air-Sea Battle. However, the relative merit of these solutions and the balance among them cannot be adequately calculated on the basis of an unexamined strategy.

Command vs. Control

The U.S. Navy has enjoyed comprehensive command of the sea since World War II. Rightly considered, this is a strength relationship between or among navies. If, because of the results of battle or other factors, one navy is sufficiently strong with respect to actual or potential opponents, it gains freedom of action in various ways, including the ability to disperse its combatants widely to exercise sea control—the protection of ships—in multiple locations.[2] In addition, the nation may disperse its capital ships to exercise command of the sea; that is, through the threat or use of power projection, enforce the rules of an international system that is congenial to its interests.[3] This latter function reflects U.S. Navy strategy since 1945. Exercising command demands the use of the most capable ships; in the U.S. case, aircraft carriers. Boasting powerful land-attack capabilities, these ships have anchored the U.S. policy of maintaining strategic stability. Command of the sea permits not only their dispersal but also the ability to move them close to the scene of action ashore and compromise their mobility, actions that in other ages would have been considered excessively risky to capital ships.

As China and Iran develop sea-denial capabilities, the risks to the carriers escalate and the traditional logic of naval strategy asserts itself. Capital ships are to be used to secure command of the sea, and this means they must be committed only when the prospect for gaining such command exists. The problem is that neither China nor Iran is looking to seize command, nor are they structuring their forces to do so. Instead, they seek, in the first instance, to disrupt and deny U.S. naval forces the ability to exercise sea control in their littoral waters and near seas. This may be for coercive purposes or to allow aggression via the sea.

Any such warfare will be about seizing or denying control in a specific sea area, thus the logic of strategy suggests that capital ships not be risked in such fights. This does not mean that capital ships are therefore no longer useful, because command of the sea or lack thereof forms the political and strategic context in the struggle for sea control. These fights always involve only a specific area of the world ocean, while command, especially in today's world, is a global condition. The U.S. Navy is inherently a global force, and regardless of the importance of a local sea-control fight, the rest of the ocean as well as future global strategic dynamics must be considered in risk calculations.

Peace vs. War: Crisis

While a surprise Japanese attack on Pearl Harbor initiated World War II in the Pacific, subsequent history suggests that a future naval war will ignite out of a smoldering crisis that features the intermingled maneuvering of opposing naval forces prior to the outbreak of hostilities. The Soviet and U.S. naval confrontation in the autumn of 1973 in the Eastern Mediterranean serves as a potential paradigm.[4] (The author was a junior attack pilot on the USS *Independence* [CV-62] during the crisis. The Soviet Fifth Eskadra had antiship missiles and a plan to use them. The U.S. 6th Fleet had little in the way of a responsive counterthreat. The carriers only had munitions appropriate for land attack, and the air wings had no antiship tactics; a function of the Navy's focus on power projection in Vietnam for so many years.)

Generally speaking, the tactical offense is dominant at sea, and the force that is able to strike effectively first is likely to attain an insurmountable advantage.[5] For this reason, naval forces strive to remain unlocated or at least untargeted. Maneuvering during crisis in constrained waters in the presence of the potential enemy tends to sacrifice any covertness. Thus naval forces in such a situation tend to be on a hair trigger. The situation is exacerbated if one or both sides have committed capital ships to the situation.

The U.S. naval strategy, as has been stated, is to distribute single carrier groups widely to cover key potential trouble spots around the Eurasian littoral. If a crisis erupts, and especially if it erupts quickly, the most likely scenario is that a single carrier group will be the first American force on scene. This was both normal and acceptable when there was no appreciable threat at sea. Given time, the U.S. Navy can and has aggregated multiple carrier groups on scene for sufficient combat power. However, in the future environment, concentrating a multi-carrier force will take longer due to fewer hulls, and the risks will escalate due to emergent denial systems. Moreover, the movement of powerful naval forces can as easily catalyze as deter, the British dispatch of the nuclear submarine HMS *Conqueror* in 1982 to the Falklands being a case in point. The Argentines accurately predicted when she would arrive from the Mediterranean, and this put them in a "now-or-never" frame of mind with regard to mounting an invasion.[6] If a lucrative target loaded with potent geopolitical symbolism is on scene, with more on the way, it could precipitate a dangerous "window-of-opportunity" mindset in the opposing government. British Prime Minister Margaret Thatcher reflected this logic when she said concerning British decision making prior to the Argentine invasion of the Falklands, "Most important perhaps is that nothing would have more reliably precipitated a full-scale invasion, if something less had been planned, than if we had started military preparations on the scale required to send an effective deterrent."[7]

One way to reduce the immediate jeopardy is for the on-scene carrier force to cut and run to get untargeted in more open waters until reinforcements arrive. However, this may effectively cede water space to the opponent—precisely what it wants. Had U.S. carrier groups run west of the Strait of Sicily in 1973 to get untargeted, the Soviets would have been left in possession of the Eastern Mediterranean, and Israel would have been isolated. The U.S. 6th Fleet might have been able to fight its way back in, but the Soviets would likely have gone to the U.N., asking for a halt in place of all forces for a cooling off, and the United States would have been in a politically weak position. Obviously, the subs would have remained on scene, but the signaling was strictly a function of surface ships.

The key characteristic of naval forces is their mobility; it allows for maneuver on a global scale and is also protective. Naval mobility is a continuous factor in peace and war, and in the way the former transitions to the latter. In confined seas and in crisis maneuvering it tends to be compromised in either an absolute or relative sense. Mobility, in both its senses, is most important for capital ships. We must recognize the possibility that strategic mobility of such ships could as easily be catalytic as a deterrent and that loss of protective mobility could have the same effect. The current U.S. maritime strategy, *A Cooperative Strategy for 21st Century Seapower* (*CS21*), asserts that preventing wars is as important as winning them. The analysis so far presented suggests that continuing to follow the pattern of aircraft carrier employment that has been used for the last six decades is increasingly un-strategic.

Credible Combat Power vs. Trip Wires

CS21 calls for concentrating "combat-credible" forces in the Middle East and in East Asia. It does not further define what is meant by combat-credible, but the logical inference is that such forces, whatever their composition, would be sufficiently capable to either defeat aggression outright or at least delay the attainment of some irretrievable military or political fait accompli. This is a rather rigorous standard, given that the U.S. Navy is operating in the littoral of a potential opponent who can bring significant land-based forces to bear, and who presumably has the initiative in terms of when to start the fight. The bar of combat credibility is raised even higher as new and more sophisticated denial systems are introduced. Moreover, whatever the peacetime credibility of forward-deployed naval forces, the pressures of crisis may significantly alter those perceptions among potential aggressors.

In naval warfare at the tactical and even at the operational level, seizing the initiative matters considerably, and thus U.S. doctrine publications are full of exhortations to do so. However, at the strategic level the picture changes. Options and political room for maneuver become critical. Having the luxury of making the second move tends to open up these things to government and military leadership. Assuming the United States does not shoot first in a situation in which deterrence fails, the question becomes what price it is willing to pay for the opportunity to have the second move.

That can range from negligible, as when some unmanned system is destroyed, to near catastrophic, if multiple aircraft carriers are put out of action. At the catastrophic level, powerful national emotions are involved, and this might take away scope for political maneuver by the President. On the other hand, loss of an unmanned system (unless perhaps it is a key satellite) is unlikely to amount to a casus belli, and here again, scope for political maneuver is limited. This logic leads us to the notion of a trip wire.

Naval forces have a long history of being used as such. British gunboats during the heyday of empire were not especially capable ships, but anyone attacking them had to face the possibility that at some point the Royal Navy would show up in force. The trip wire must be such that attacking it would plausibly justify a strong response, but its loss would not materially affect the fighting strength of the Fleet. On this basis alone, the LCS would appear to be a useful presence platform in constrained seas; it has a small crew and does not represent a critical combat capability. However, our analysis must not stop there.

The invasion of Iraq notwithstanding, the United States is generally not looking for a fight. Deterrence has always been the watchword and as previously stated, *CS21* says that prevention of wars is as important as winning them. On this account the concept of a trip wire becomes suspect, and the idea of credible combat power gains importance. Forces on scene in peacetime and during a crisis must have the ability to disrupt enemy aggression or otherwise inflict operationally or strategically significant harm, and they must be able to do it after riding out a first strike.

This suggests the distribution of offensive power—missiles, most easily—among numerous platforms and making those platforms hard to target. In such a situation a potential aggressor is faced with a dilemma in which the prospects for a disabling first strike are dim, but any attempt will break a trip wire and also precipitate an immediate damaging response. Meanwhile, the U.S. national command authorities hold the political cards and can dispatch, if they wish, powerful forces that can conduct a structured rollback campaign of their choosing. In this kind of scenario it would be hard for the aggressor to claim any kind of moral victory. This logic would not be lost among foreign leadership, and deterrence would be thereby improved, assuming there was any chance of its working in the first place.

"Being There"

"Being there" matters in strategically important ways. American naval forces have just and sufficient reason to operate continuously in the littorals of Eurasia in light of historical developments since World War II and as a matter of global public good as outlined in *CS21*. The strategic mobility of naval forces is a massive advantage, but for reasons outlined here, it can be a double-edged sword in modern conditions. The U.S. Navy would be well served to create new kinds of forward forces that are less strategically mobile but more tactically suitable for the operational conditions that are emerging in the Persian Gulf, South China Sea and elsewhere. The LCS may fulfill part of this requirement, and thus it is entirely appropriate that a number of them are being stationed forward. However, the LCS plus existing strike-group ships do not appear to fully answer the new requirements.

Since the age of fighting sail, confined seas, estuaries, lakes, rivers, and bays have been the province of flotillas—groups of small, numerous, and often specialized craft. The range of modern weapons has expanded the breadth and extent of the areas that should rightly fall into the flotilla domain. Through flotilla operations the U.S. Navy can retain a strategy of ceding no water space, either as a matter of peacetime policy or crisis maneuver.

Although the U.S. Navy has not often used flotillas (Vietnam riverine operations come to mind), they are a legitimate element of naval strategy and have their own singular set of operating principles—most of which the Navy has forgotten. Sorting out ship, boat, weapon, and systems designs is only partially a matter of technological push. There should also be a strategic and operational pull that defines missions. This will establish suitability criteria for various designs. Unfortunately, the Navy has not yet adopted the strategic concept of flotilla operations. It will therefore continue to throw money at developing stronger defenses for capital ships so that they can be used in strategically inappropriate ways.

Flotilla operations should be called out in the next refresh of *CS21* and incorporated into the Naval Operational Concept. They should be gamed out thoroughly enough that capability requirements and criteria can be established. Working top-down in this manner will guide and facilitate bottom-up technical development and eventually tactical innovation. In this way the Navy can make its investments of scarce resources with more confidence. A program office for flotilla operations should be established, such as was done for Air-Sea Battle and irregular warfare. It should be noted at this point that flotilla operations would involve elements of irregular warfare and might contribute to Air-Sea Battle, but are best approached as a distinct branch of naval warfare. Operationally there should be a flotilla task-force commander reporting to the regional Navy component. Flotilla operations will be regionally specialized and may require tailored forces with distinct characteristics as well as doctrine customized for the particular circumstances of the region.

There are some cultural and emotional Rubicons for the Navy to cross in order to achieve effective flotilla operations. It must embrace truly small combatants, and the surface-warfare community must not orphan flotilla sailors and officers; they must have viable career paths. Flotilla operations as envisioned here and later in this issue will be forward and closely connected to local navies and nations. Thus there seems to be a natural link between a putative flotilla specialty and the Foreign Area Officer program. The Navy Expeditionary Combat Command may be spliced into it, and it stands to reason that Special Forces would also be involved.

Flotilla operations are a feasible and appropriate response to the intersection of emerging access-denial technologies, the logic of naval strategy, and the geopolitical imperatives of the United States that require it to "cede no water" to those who would seek to limit freedom of navigation or conduct aggression at sea.

Notes

1. Milan Vego, "Go Smaller; Time for the Navy to Get Serious About the Littorals," *Armed Forces Journal* (April 2013), www.armedforcesjournal.com/2013/04/13617202.
2. Julian S. Corbett, *Some Principles of Maritime Strategy* (London: Longmans, Grean and Company, 1918), Part II, Chapter II, 100–04.
3. Robert C. Rubel, "Command of the Sea: An Old Concept Resurfaces in a New Form," *Naval War College Review* (Autumn 2012): 21–33. This perspective on command of the sea was derived from George Modelski and William Thompson, *Seapower in Global Politics 1494–1993* (Seattle: University of Washington Press, 1988), 16–17.
4. For additional insight, see Lyle Goldstein and Yuri Zhukov, "Tale of Two Fleets: A Russian Perspective on the 1973 Naval Standoff in the Mediterranean," *Naval War College Review* (Spring 2004): 27–63, www.usnwc.edu/Publications/Naval-War-College-Review/2004---Spring.aspx.
5. Wayne Hughes, *Fleet Tactics: Theory and Practice* (Annapolis, MD: Naval Institute Press, 1986), 25.
6. Lawrence Freedman and Virginia Gamba-Stonehouse, *Signals of War* (Princeton, NJ: Princeton University Press, 1991), 65–78.
7. Margaret Thatcher, *The Downing Street Years* (New York: HarperCollins, 1993), 174.

11 "ON MARITIME STRATEGY"

CAPT J. C. Wylie Jr., USN

Contemporary Naval War College students will recognize the "case study" method that Captain Wylie employs in this prize-winning essay (Honorable Mention in the Naval Institute's annual Prize Essay Contest—his second such award). He was serving his second tour of duty at the College when he penned this instructive treatise.

More than a decade later, Wylie's *Military Strategy: A Theory of Power Control* was published, followed in 1989 by a revised Naval Institute Press edition with an introduction by famed War College professor John Hattendorf as part of the Classics of Sea Power series. Professor Hattendorf described Wylie as "a rarity among American naval officers ... the first serving officer since Luce and Mahan ... to become known for writing about military and naval theory."

This early article contains some of the theoretical underpinnings of his later work, contextualized in the early Cold War milieu. At the very heart of his thinking, then and later, is the postulate that "the aim of any war is to establish some measure of control over the enemy." What makes this article particularly useful is the application of that and other ideas to the maritime environment.

"ON MARITIME STRATEGY"

By CAPT J. C. Wylie Jr., USN, U.S. Naval Institute *Proceedings* (May 1953): 467–77.

Over the last half century there have been two generally accepted approaches to the study of maritime strategy. The first has been an analysis of the component elements of maritime strength, with Mahan's classification of these as geography, naval power, merchant marine, and the like usually serving as the basis for this kind of study. The second approach, more prevalent in our generation, is the discussion of strategy in terms of specific types of operations such as fast carrier strikes, anti-submarine warfare, or organized overseas transport. I think both of these avenues of approach tend to obscure, to some extent, the coherent form of the basic strategy that lies between these two, the strategy that grows from the components to give continuity and direction to the operations. It is this middle ground that I shall explore, the area in which a basic element of strength is transformed by an idea into a positive action. It is a sailor's concept of strategy, what it is, how it works, to what end it is followed, and what its problems are.

In doing this I shall present the subject from those four aspects that seem to me necessary for an adequately rounded appreciation both of the underlying idea and of the translation of that idea into practical and useful results. These four aspects are, first, a theory of maritime strategy; second, past experience in the use of maritime strategy; third, some of the factors that complicate its use in our time; and fourth, our contemporary use of military power and the tendencies with respect to maritime strategy.

I

A Theory of Maritime Strategy

The aim of any war is to establish some measure of control over the enemy. The pattern of action by which this control is sought is the strategy of the war. There are many types and levels of strategies, and many ways in which they may be classified. But since the subject of this discussion is maritime strategy, the classifications we shall use have already been determined. The three main streams of strategic thought in this sense are maritime strategy, continental strategy, and, more recently on the scene, air strategy.

Here, at the beginning of this discussion, it should be emphasized that clear-cut separations are artificial. In practice there is, and must be, a good deal of overlap and merging; the strategies are deliberately set apart from each other in this treatment of the subject only for purposes of study and analysis.

I use the term continental strategy to indicate a pattern of employment of armed forces in which the major and critical part of the action to establish control over the enemy is directed against his armies along a central land axis. All other efforts are in support of the central drive of this continental strategy. In spite of the descriptive title that I have elected to use for this type of strategy, the involvement of an entire continent is not necessarily implied.

The term air strategy I use to indicate an over-all war strategy in which the decision is sought primarily by air action, with predominant emphasis on strategic bombardment. All other efforts are to a greater or lesser degree subordinate to that.

A maritime strategy is one in which the world's maritime communications systems are exploited as the main avenues by way of which strength may be applied to establish control over one's enemies.

Maritime strategy normally consists of two major phases. The first, and it must be first, is the establishment of control of the sea. After an adequate control of the sea is gained comes the second phase, the exploitation of that control by projection of power into one or more selected critical areas of decision on the land.

Too often the first or blue-water phase of maritime strategy is regarded as the whole process rather than no more than the necessary first half. Most naval history, for example, concerns itself with the struggles for control of the sea, the naval battles, the protection of commerce, and the blockade in one form or another.

This phase, also, is the one that attracts the greater part of our own professional attention, and it is the phase that most landsmen accept as the entire concept when one introduces the subject of maritime strategy.

Within the first phase, the control-of-the sea segment of the over-all pattern, there are initially two components of control. They will be considered separately for analytic purposes, although the actual conduct in war may so closely interweave them that the separate goals may not be superficially apparent. One of these components is ensuring one's own use of the sea; the other is denial to the enemy of his use of the sea. At least in the early stages of the struggle for control, these two goals can be analytically considered more or less apart from each other; and, as long as neither contestant predominates in both, there remains a fairly clear delineation of these two functional components of the struggle for control. Not until one sea power emerges as dominant in both these components do the two of them merge into the single problem of positive and willful control of all that moves at sea. This is the ideal condition that has many times been striven for but never, except possibly at the very end of the War in the Pacific, been attained in near perfection.

The naval strategies (as differentiated from the more inclusive maritime strategies) and the naval tactics of both contestants are largely determined by the status and progress of the struggle for control.

If, for instance, two nations having roughly equal naval strength go to war, the attention of each of them must be devoted primarily to the fight for naval supremacy. This type of naval contest has a dual purpose in that it embraces both components of control of the sea—the positive aim of securing one's own use of the sea (by destroying the force that could hazard that use), and the negative aim of denying to the enemy his use of the sea (destroying the force that could protect him). This was the case in the recent War in the Pacific when the American and Japanese fleets fought it out until we finally established an ocean-wide control adequate to our needs.

A somewhat different situation exists when two opposing nations start with unequal naval strength. The primary aim of the stronger in this situation is to protect and extend its own use of maritime communications by both passive and active means—passive in defending its own forces at sea, and active in seeking out and destroying enemy forces that offer threat to the use of the sea by the stronger power. The primary aim of the weaker is to interfere with the stronger's use of the sea by resort to some specialized technique, such as a war of attrition, a deliberate hoarding of naval threat, or an attack on the stronger's commerce. This was the case in the last two wars in the Atlantic when Germany, as the weaker, hoarded her major combatant ships and placed her hopes in her attacks on allied commerce.

There are many refinements to be applied to this theoretical outline in the actual struggle for control of the sea. Particularly in the early part of a war, control is normally in dispute with neither antagonist able to utilize the sea to his own satisfaction. This dispute leads to two other situations which are frequently encountered; one is local control of some part of the sea, and the other is temporary control. The two are frequently combined as when the Germans attained a temporary local control of the waters off Norway long enough to permit their invasion and consolidation of that position. The control of the Mediterranean was in dispute for the first three years of the recent war, with both sides at times having temporary control of the Central Mediterranean, while the British, except for one indecisive moment, never lost local control at either end.

This is an all too brief outline of the problem of control of the sea, the necessary first phase of a maritime strategy. When a maritime power is reasonably successful in securing the sea to its own use (that is, in repressing the enemy's power to interfere unduly), then it can turn to the second, or exploitation, phase of maritime strategy. And here the subject becomes considerably more slippery, which is really not surprising since it is, in actual fact, a far more subtle proposition than most of us initially realize.

In order to discuss the exploitation of sea power it is necessary to return to the premise that opened this discussion, the assertion that "the aim of any war is to establish some measure of control over the enemy." If this premise is accepted (and its acceptance in general substance is critical to this theory of warfare), then the next step is the examination of methods of establishing control.

In wars between powers having their major strength in ground forces, the defeat of one of the two contending armies has generally led to victory. This has been the situation when two continental strategies have opposed each other. In wars in which at least one of the two contestants was a major sea power, the defeat of one contending navy and the establishment by the other of control of the sea has generally led to victory. But this victory has been reached only when the dominant sea power has exploited his strength at sea by projecting at least one other element of force to establish control over the enemy on the land.

In some cases the strength at sea has enabled the naval victor to launch a ground force into a critical part of the other's territory. In these instances the soldiery has been the direct instrument of control that clinched the issue at hand. The sea-borne invasions of Sicily and Italy a decade ago were the exploitation by ground forces of the naval seizure of control of the sea. We have already noted the sea-and-air-borne invasion of Norway by the Germans. This was an exploitation, using ground forces, by a nation with temporary control of the local sea area.

Another means of exploitation by sea power is the use of economic force for the application of control. When the Armada was defeated, England intensified her blockage of Spanish trade with the New World and eventually choked Spain almost to death. Spain has never recovered. In the Anglo-Dutch Wars, England established her naval control at sea and was in position to clamp down on Dutch commerce as she had done against the Spanish. But the Dutch saw what was in prospect, acknowledged the potential strength of British control (augmented by Britain's geographic advantage), and reached agreement with Britain before having to undergo the painful process of having all Dutch commerce destroyed. In the case of the recent war with Japan, the advancing American control of the sea was exploited to stifle Japanese overseas communications. Economic suffocation was the primary instrument that enforced Japanese acquiescence. She was dependent on sea communications, not only for her existence as a major power, but for her very life.

There are a few instances in which the instrument of control has been some other force, a political pressure of this-for-that, a direct or indirect bribery of men having power of decision, or a revolt somewhere inside the structure of the enemy power. But the main methods by which force has been applied to establish control over the enemy have been these three: a victory by the armies of one land power over another; a victory by a sea power exploiting her power at sea to project a frequently smaller but strategically decisive ground force for the actual establishment of control on land; and a victory by a sea power exploiting her power at sea to project an economic force toward the eventual establishment of governing control over the enemy in his own land. It is the second and third of these, the two main methods of exploiting power at sea, that form the basis of the second phase of maritime strategy.

It should be noted that, in practice, the exploitation of sea power is usually a combination of general slow stiflings with a few critical thrusts. These latter are frequently spectacular and draw our attention to the exclusion of the former, while in point of fact the critical thrusts would not be critical were it not for the tedious and constant tightening of the screws that makes them possible.

II

An Illustration of Maritime Strategy in Use

Up to this point I have outlined the basic pattern of action from which a maritime strategy may be compounded, never in pure form, of course, but with the appropriate blending of armies and navies and air forces, and of political and economic and psychological forces.

Now, after this theoretic description of maritime strategy, let us examine a specific problem. How might we evolve a strategy for the United States and how would we judge its validity today?

I think the soundest way to reach toward answers in any such inquiry is first to turn to comparable historical experience, to recognize the points both of similarity and of difference in the two situations, and then to take advantage of that experience in light of our own specific circumstances.

The recent war with Japan is already accepted as the modern naval classic. But the problems faced in a war with insular Japan and the problems faced in a possible war with a continental great power are not the same. I do not think the War in the Pacific is a valid precedent to turn to for a study of maritime strategy relevant to a war with a power whose base is Eurasia. Correspondingly, I do believe that much of the confusion in our present naval thinking is the result of trying, without careful discrimination, to adapt the war with Japan to a prospective war with a great land power.

Let us look at our situation. We are a great sea power, geographically set apart from the continent by intervening waters. Our hypothetical opponent, a great land power with a much smaller sea power potential, is firmly in control of much of Europe and is seriously threatening the rest. Has that situation ever existed before? And how was it managed?

Yes, that situation has existed before, several times, under reasonably similar circumstances. It existed and was managed with a fair degree of similarity in the First World War and again in the Second World War. But it was managed in what I think was an even closer analogy a hundred and fifty years ago. I have selected that third one as our point of departure in this discussion because it illustrates the application of a maritime strategy with fewer obscuring complexities than either of the two more recent situations. I shall outline the experience of Britain in defeating Napoleon and then, after the skeleton of that strategic process is exposed to view, superimpose on it some of the complicating factors that confront the strategist today.

The British struggle with Napoleon illustrates quite clearly the two major phases of a maritime strategy. At the start of the war, late in the eighteenth century, both Britain and France had a major strength at sea. The struggle initially was a struggle between fleets for control of the sea, a control that was in dispute for many years. This is "the portion of maritime history that most of us are familiar with, and this is the first phase that was completed with the successful campaign culminating in the victory at Trafalgar.

The second phase of that great war is remarkably unappreciated. It opened with Britain's having gained control of the sea at Trafalgar, and it is here that I am going to draw the comparison with our situation today. We have the potential, if not the actual strength, to establish a workable control of the sea. It need scarcely be said that this will take considerable doing, but I do not question that we can gain control of the sea when we need it.

So let us start from that point in 1805, well along in the middle of the war. At that time Britain, as a major sea power, found herself facing the problem of how to defeat France, a major land power, firmly in control of much of Europe and threatening the rest.

The ten years from Trafalgar to the final downfall of Napoleon in 1815 present, at first glance, a very confusing picture. Over all the scene lay the shadow of the seemingly irresistible and enduring strength of the Emperor's armies. There was a succession of apparently disconnected battles from one end of Europe to the other. There was a continuing bitter economic warfare that reached its climax with the Berlin and Milan Decrees by which Napoleon tried to exclude Britain from her markets. These were met with the Orders in Council by which Britain attempted to control and limit the commerce of Europe to her own advantage. There were unsteady and changing governments, now opposed and now subordinate to Napoleon. There was propaganda, intrigue, bribery, and treachery. And all through that period there were the tremendous British grants of monies to potential allies all over Europe; indeed, Britain literally financed most of Europe at one time or another during those turbulent ten years of war. When examined in terms of these details, it seems almost incredible that Britain ever won. But when the entire period is taken under scrutiny, then three fundamental factors, superimposed on the basic and continued maintenance of her control of the sea, emerge to give coherence to the actions over those years.

First, Britain, in the exploitation of her maritime strength, never slackened her pressure on the French all around the periphery of the Empire. The economic war was waged bitterly and continuously, and advantage was taken of every economic strain that developed within the continental system.

Second, Britain, in exploitation of her sea communications, never missed an opportunity to launch an army against a vulnerable point in Napoleon's armed strength. Whenever the Emperor moved one way, then Britain and whoever happened to be her allies at the moment stabbed from the other. In Portugal, in Spain, in Austria, in the Low Countries, in the Baltic, Napoleon was never secure. Whenever Napoleon managed to counter these threats with his own greater force, then Britain took her profits, cut her losses, and withdrew, biding her time till opportunity showed again.

Illustrative of the pressures on Napoleon from the sea were the concurrent activities of two British commanders, the Duke of Wellington in the Iberian Peninsula and Admiral Sir James Saumarez in the Baltic. In 1811 while Wellesley was producing what Napoleon described as the "Spanish ulcer," Saumarez, commanding the dominant sea strength along northern Europe, brought about some unpublicized but critically important secret meetings on board his flagship. In these he induced agreements with Sweden and Russia in which the Czar was given the military and political freedom that he needed to turn on Napoleon. This soon brought Napoleon's armies from Spain into Russia, and that wholesale calamity in 1812 needs no comment except to note that it could never have happened had not British sea power been applied with remarkable political agility in Napoleon's rear.

Third, in the exploitation of her sea power Britain never did formulate and commit herself to a single military plan by which she expected to win the war. She never lost sight that her goal was the defeat of Napoleon; she never missed a chance to apply pressure where it hurt; but there was no constricting rigidity of plan nor any premature commitment in her strategy. Basic to her maritime concept was her practice of taking timely advantage of opportunity as it opened to her.

In our own contemporary atmosphere of intensive and inclusive planning, we might pause to realize anew this peculiar advantage in exploitation of sea power. It is the capacity to manipulate the placement, the timing, and, in great measure, the weight of the strategic centers of gravity on the land.

Britain had the ability and the will, and she exploited to the fullest her control of the maritime communications systems of the world. Operating from the base of her firm control at sea, Britain and her allies continued their penetration of every crevice in Napoleon's armor until finally his structure fell at his heels. Napoleon himself seems never to have realized that it was the ubiquity of Britain's sea power that lent the repeatedly resurgent and finally victorious strength in the defeating of Napoleon.

How truly remarkable is the similarity between today and a century and a half ago. The shadow of the dictator's army over the unwilling peoples of Europe. Their hope of independence to be regained with the help of the sea power from off the continent. The Berlin Decrees in the one case and the Iron Curtain in the other, and the intense efforts in both to build an independent economy behind these barriers. The pulling and hauling in the formation of alliances with the great maritime strength of the day. The struggling free nations of Europe helped to their feet by the financial and economic strength of the power across the sea. And the clear understanding by these two great maritime nations, then Great Britain and now the United States, that Europe must be kept free of dominance by a single power if they would themselves survive.

III

Factors That Complicate Modern Strategies

These are the similarities in the two situations today and a hundred and fifty years ago. But striking as they are we cannot disregard the fact that intervening between that time and this are the tremendous upheavals of the industrial revolution, its contemporary successor the technological revolution, and the continuing social and political revolution that still surges throughout the world. These have changed the tools and techniques of warfare almost beyond recognition. In many important respects the visible and active appearance of warfare bears little or no resemblance to that of a century and a half ago. But more subtle than these obvious changes in combat activities are the problems of whether and how the modern innovations have altered the underlying patterns of war, the basic strategies of war. While many of the skills of men-of-warsmen today bear little relation to those of the men who sailed under Collingwood, can we properly infer that the strategic problems that faced Collingwood and Nelson and Barham and Pitt are equally unrelated to those facing their successors today? That question is the one at hand when we set ourselves to judge the value of yesterday's experience in today's situation.

In order to open up that question I have selected six major complications of warfare that have grown out of the industrial revolution to perplex the strategist today, six problems that either did not exist before or have undergone such marked change as to be in fact new problems. These are: mechanization in war, explosives in war, arms and revolution, logistics in war, the phenomenon of flight, and nuclear energy in war. This is certainly not an exhaustive list, but it is, I think, an indicative one. The extent to which these matters alter the fundamental pattern of action by which we seek to establish control in war is the extent to which we must modify yesterday's experience in applying it to today's situation.

MECHANIZATION IN WAR

When we consider the industrial revolution, we realize at once the unbelievable progress that has given us tanks and jeeps and steam vessels and submarines and automatic weapons. The difference between a primitive or man-power armed force and an industrial or mechanized armed force is apparent. But all the problems growing from these differences have not yet been generally recognized. Our military attention has been concentrated almost exclusively on the problem of fighting one mechanized armed force with another mechanized armed force. That is true in all three of the services, Army, Navy, and Air Force. However, there are two other problems to be considered. One is the business of devising strategies and tactics for use by relatively primitive armed forces against highly mechanized ones; the other, devising strategies and tactics for use by mechanized forces against primitive forces. That latter problem is a very real one. It faces us today in Korea, and I think we have failed to recognize it as basic. It is a direct challenge to the validity of the strategic concepts applied in Korea. It would be a challenge in a much greater field if the war were to widen.

Our present weapons and techniques are the best we can devise for use against armed forces such as our own. The question we must ask is: are our present strategic concepts, techniques, and weapons also the best that we can devise for use against armed forces whose primary strength is manpower rather than highly refined and complicated machine power?

With respect to ground forces the importance of this query may not be too critical as long as the infantry remains recognized as the focal point of ground strength. That should insure maintenance of the man-power perspective no matter how much machinery may be involved.

With respect to naval forces, a careful pondering of this question could, I believe, lead to a shift of emphasis in our fleet preparations from the blue-water reaches of the sea to the inshore soundings. Apart from countering the atypical (though very real) hazard of uniquely efficient submarines, I believe that a large proportion of our naval effort, particularly in the exploitation phase of a next war, must be put into tools and techniques that can seize and exploit control of the shoal and restricted waters along the enemy littoral and penetrating into the enemy territory. This subject deserves elaboration, but there has yet been no satisfactory statement of the problem, much less a satisfactory approach toward its solution. That a problem does exist, that it will require a fairly large change in prevailing strategic concepts, and that it will require the evolution of basically simple tools and techniques not now at hand, I am sure. But no one has yet been able to suggest the shape of a generally valid concept tailored to this need, nor the particular functions of the tools that we must adapt or devise. The problem concerns the maintenance and exploitation of control on inshore waters, a matter that I think was handled better a century and a half ago than it is today.

With respect to air forces, the problem takes a somewhat different turn. While there is no such thing as a primitive or man-power air force, we do find ourselves faced with the business of fighting a relatively primitive force with a highly mechanized air force. It is a problem that must be faced by naval aviation as well as by independent air forces. Here, more than anywhere else, we have fallen into the trap of casting the enemy in our own image. To use a specific illustration: we have done all our planning of air interdiction on the assumption that, if the interdicting effort is strong enough, it will succeed. Against highly mechanized ground forces this may well be true; such ground forces are a most susceptible target. Against a piggy-back army, one whose basic reliance is in men and animals, I think this assumption is not valid. The point is that the theory of interdiction, air against ground, must be modified to the extent that the possible effectiveness of interdiction is a function in part of the strength of the interdicting force and in equal part of the susceptibility of the target to interdiction. A highly mechanized target is maximally vulnerable; as the target becomes more primitive, the susceptibility approaches zero.

That is one type of modern complication that the strategist, maritime or otherwise, must consider.

EXPLOSIVES IN WAR

Now let us take up a different aspect of the industrial revolution. Gradually, over the last century, the function of explosives in warfare has changed, and I think we have missed the significance of this change.

Originally the explosives, gunpowder in one form or another, were used primarily as a propellant for missiles. The purpose of these missiles was direct destruction—the direct killing of men or the sinking of ships. This led to an efficient imposition of control with little or none of what I shall describe as "over-kill." This word I use to indicate that proportion of effort which cannot be used for the direct establishment of control.

Today the infantry rifle is the remnant of this once universal method of warfare. We now use explosives as a propellant as we did before. But we also use explosives both as an agent of destruction at the target and as the on-target propellant of secondary missiles of destruction. The nature of contemporary industrialized and mechanized targets, both civilian and military, invites the use of explosives against them on a grand scale. Modern methods encourage the use of explosives as a general agent of destruction.

A result of this has been a prevailing tendency to equate destruction with war, and this in turn leads us to associate the idea of maximum destruction with victory.

In this partly justified and partly superficial thinking, there is a fallacy. That fallacy is our forgetting that the purpose of destruction in war must be the achievement of control. Other than that it has no point. The degree to which destruction contributes to control is the degree to which it contributes to final victory. Destruction by the massive use of explosives carries with itself the inherent characteristic of a large proportion of over-kill (with its very important secondary effects) and thus a lessened proportion of direct control.

The relationship between destruction and control in war is one critical measure of the efficiency of the conduct of war.

The maritime strategist has long been aware of this, his appreciation emphasized by the comparative economy forced on him by the nature of the tools with which he works. The essence of the exploitation of sea power is the projection of concentrated power to critical points of decision, the establishment of a maximum of control with a minimum of war's general destruction. From this has grown the sailor's firm belief in the need of peculiarly specialized types of ground strength and of air strength as built-in components of maritime strength in order that he may impose his decisive control at critical points of his own choosing. This is a compelling reason why marines and aviators are integral units of the naval service.

ARMS AND REVOLUTION

In addition to mechanization in war and the role of explosives in war, there is a third facet of the industrial revolution in war that, in a different field, modifies somewhat the classic patterns of strategy.

During the Napoleonic Wars and until fairly recently, it has been possible for any determined people to revolt almost at will. Before mechanization it was a relatively simple matter for any dissident group to lay its hands on the necessary tools of war and revolution. Some pikes and halberds could be improvised; smooth-bores and even rifles could be made or stolen and stored locally until the time came for their use. But the tools of warfare have grown so complex and expensive within the last two generations that in a modern society only the state itself can organize and pay for the production of arms. This means that the support of an army, with its arms, is now a necessary ingredient of revolution. Unless a state's own army joins the rebels, then the help of an outside army must be directly available before any revolution can be successful.

This concerns air strategy to whatever extent airborne and air-supported troops may be adequate to the needs, though even this is more a matter of ground force than of air force interest. More fundamentally, it concerns continental and maritime strategy. An outright revolt within the enemy's political or military structure must not be encouraged until an adequate and sympathetic ground force is directly at hand to support it. This limits the possible areas of revolt in war to those along a continental front or along the accessible littoral behind the front. Since this restriction on revolution has developed, a mobile sea-borne and sea-supported ground force has become increasingly important to the exploitation of this type of potential weakness in an enemy.

LOGISTICS IN WAR

Directly related to the growth of the industrial and technological revolutions is the problem of modern logistic support. Primitive armies could, and to a large extent still can, live off the country. Mechanized armies cannot. A fleet under sail could stay at sea almost indefinitely; indeed, Nelson kept his Toulon blockade for over two years without once leaving his flagship. A modern navy can stay at sea for considerable time, to be sure, but not without enormous effort in logistic support. Air forces, while there is no preindustrial comparison, are by their nature the most logistically helpless element of armed force. In all types of strategies, continental, maritime, or air, the logistic factor must weigh heavily in arriving at decisions, both with respect to the quality and quantity of material needed and with respect to the time and cost required for its delivery.

It is of interest to compare the three basic types of strategies in the matter of logistic vulnerability. In the continental strategy, the mechanized army is far more vulnerable than its predecessor by reason of its logistic dependence. While it is tactically more mobile, it is strategically an infinitely more ponderous mass to move or to redirect. In a strategy basically maritime, the bulk and complexity of logistic support is incalculably greater than that of the classic sea powers, but the application of logistic support may actually be a good bit easier. Easier not only compared to primitive maritime force but compared to mechanized continental force. The flexibility of contemporary maritime communications systems compared to those on land, and the lesser degree to which they can be critically interrupted after control is established at sea, combine to make the exploitation phase of a maritime strategy quite attractive when balanced against a drive toward a similar goal by over-land avenues.

After indicating the scope of the industrial revolution's logistic effect on both continental and maritime strategies we can see that, while it is complicating, it is not unique. The problems involved are not novel; they are distorted and magnified, but they do not invalidate the traditionally accepted bases either of the continental or of the maritime strategies.

With respect to air strategy—and here I am going to merge logistics with the next major topic, that of flight—the logistic effect of the industrial revolution takes a somewhat different turn. The logistic problems introduced by the industrial revolution are the basis of prevailing air power theory. The theory of strategic bombardment and the theory of interdiction are both predicated on an assumption of critical vulnerability of the enemy's logistic support. In comparing the capacities of continental or maritime strategies with air strategy, or in the weighing of any derivative lesser strategies, the first point of examination should be this: to what degree is the assumption of the enemy's logistic vulnerability valid? The continental or maritime strategies are not completely dependent on this or on any other one assumption; the air strategy is. Only to the degree that this assumption of critical vulnerability is accepted can a comparison be continued past this initial point. Only to this degree can we then make inquiry as to the relation between logistic destruction and the achievement of strategic control. Only within these limits can judgments be valid.

FLIGHT

Quite apart from the logistic base of the specialized theory of air power, the phenomenon of flight has had three generally recognized effects on warfare.

First, it has extended the range and quality of observation in the conduct of war, enough so that both the tactics and the strategies of war have been affected. This change in the range and quality of observation has probably affected naval warfare more than war on the land. Flight has had more influence than any other factor on the management of the age-old problem of the unlocated enemy.

Second, flight has extended the range and affected the use of destruction in war and altered the comparative value of targets of destruction. The relative importance, for instance, of cities in warfare, now that they have become industrial centers of power, has undergone quite a change since the coming of the airplane.

Third, flight has introduced a new capacity for transportation, a capacity whose capabilities and limitations are so well understood that they need not be detailed here.

These three effects of flight—the changes in observation, destruction, and transportation in war—have not lacked for attention in military thought.

Finally, flight has introduced the proposition that there exists another great basis of strategic thought, that is, air power as distinguished from sea power and land power. Needless to say this proposition has not been universally accepted, and the skeletal frame of dispute with respect to air power theory has not yet been made clear. Until that is done, there can be no general acceptance or rejection of the theories of air strategy, and that lack of general acceptance or rejection is the point I wish to make. I believe a very real influence on strategic decision in any military or naval problem is created, not only by the obvious existence of flight, but by the uncertainty stemming from efforts to fit it into its proper and accepted place with corresponding military and naval activity. The maritime strategist must adapt his practices not only to the physical fact of flight but to the psychological fact of uncertainty as to its niche in the military power complex.

NUCLEAR ENERGY IN WAR

A direct result of the technological revolution is nuclear energy. We have already experienced its logarithmic increase in capacity for destruction. We are beginning to see a comparable increase in capacity for movement. We seem to be fairly well beyond the emotional shock associated with its initial display. And we are acquiring acceptably objective information as to the capabilities of nuclear weapons in terms of direct destructive effects.

It seems to me that the primary unsolved problem in the field of military employment of nuclear weapons is the problem of explosives and their over-kill—the relationship between destruction and control that has already been introduced into the discussion.

In tactical terms the results are probably calculable. Against military targets on the land or the sea, the effect of atomic bombs will be to force a revision of the pre-atomic techniques. Against non-military targets, the imponderables decidedly complicate the issues. I believe the availability in fair quantity of nuclear weapons will force us either to re-examine our notions as to what may be acceptable results of war, or to re-examine our apparent intentions with respect to their employment. This problem, of course, is closely related to the one we face in appraising the position of aviation in the total military power structure.

The maritime strategist, I think, is fortunate in that the nature of his strategic theory does not induce an almost inevitable dependence on the use of nuclear weapons against non-military targets. So much, in those cases, is beyond calculation. Success is dependent to a governing degree not on what we do but on what the enemy does. We cannot accurately predict enemy behavior, and thus we must gamble on how an enemy will react to the side effects of the considerable over-kill inherent in the use of nuclear weapons. That is a most difficult hurdle to overcome when one's goal is recognized not as the delivery of destruction but the establishment of control over the enemy.

IV
The Pattern of Strategy Today

These six problems introduced into warfare within the last few generations are all of major importance. All of them, in one way or another, have appeared to "revolutionize" strategy. Certainly each of them has markedly altered the climate in which the strategist operates and has modified the techniques with which he puts his strategy to practice. But none of them, in my belief, has yet demonstrated conclusively that it has changed the basic patterns of strategy. These problems and others like them are still in the process of digestion in all phases of warfare. They are problems that can be resolved and, in most cases, are being resolved in practical application. The capacity of a maritime strategy to adapt itself to these major changes is one of the reasons why I believe a maritime strategy should be a most attractive one to the United States in her present situation. Let us see how it is working in practice today.

As early as 1946 the United States became aware that there was a very real possibility of all of Europe's falling under the domination of a single great power. There were different interpretations of the type of hazard that this situation would present—military, political, social, economic, or ideological hazard—but these need not concern us. We may start from the point where a hazard was recognized and trace our action from there.

Greece and Turkey were both under pressure by Soviet Russia. It was to the interest of the United States to prevent communist domination in those two countries. They were given military and economic assistance by the United States in sufficient strength to offset the communist pressure. This is most interesting because of the geographic situation involved. One of these countries has a land border common to Russia; the other, a land border common to a Russian satellite. Both of them are about five thousand miles from the United States.

But both of them are accessible by sea. This situation gives rise to the astonishing paradox: Greece and Turkey are closer to the United States, in political and economic and military terms, than they are to Russia. The common frontier of the sea and our exploitation of maritime communications systems make these two countries more accessible to us than to the communists.

During the late 1940s several of the Central European nations tried to stay out, or break out, of the Russian orbit. Poland, Czechoslovakia, Rumania, Hungary, and Bulgaria succumbed. Only Jugoslavia succeeded in breaking out of the Iron Curtain. Of all these countries, only Jugoslavia had access to a sea under Western control. I think that fact is significant; and I also think that if we had had control of the Baltic we would not have lost Poland.

Later NATO was formed. Many men fail to realize that this North Atlantic Treaty Organization is, by its very name, an alliance of maritime nations. The common bond in NATO is the bond of the maritime communications system centered in the North Atlantic. It is significant that Turkey, at the far end of the Mediterranean, which we control, is a member of NATO, while Sweden, at the very entrance to the Baltic, which we do not control, is not a member of NATO.

In the early days of NATO, a military organization was started for the immediate purpose of insuring the survival of the Western nations on the continent. The structure of this organization indicated that it was designed for immediate defense against the direct military hazard of the continental strategy opposed to it. Since that time the NATO organization has been filled out. The Supreme Allied Commander in Europe is properly an army commander. His Commander in Chief, North, is functionally and properly a naval commander. His Commander in Chief, South, should be for the same reason a naval commander. The Supreme Allied Commander Atlantic, co-equal with the Supreme Allied Commander Europe, is, as he must be, a naval commander. This present organization means that the United States and her colleagues in alliance clearly recognize the value of a strategy whose governing element is control of the maritime communications systems.

Let us compare the implications of this command organization with the elements of a maritime strategy that we identified at the beginning of this discussion. The first phase would be to establish control of the sea. The Supreme

Allied Commander Atlantic and the two subordinate naval commanders of SACEur, the CinCNorth and the CinCSouth, are organizationally situated to insure that control. The second phase would be the exploitation of sea power. The two commanders on the north and south of Europe not only command naval forces, but they command the needed associated ground and air forces to exploit the control of the sea that they establish. The Commander in Chief Atlantic is in position, not only to insure reliable communications and support for his opposite number on the continent, but also to apply the power of his maritime strength either directly to Western Europe or through the sea on either flank by way of the commanders-in-chief in the north and south. The Sixth Fleet, for instance, is basically an Atlantic unit potentially applied at present through the CinCSouth.

Vast though it be, this is only a portion of the total picture. NATO does not include all of the United States' interests, nor does it include all of the British interests, in potential wars all around the globe. These two nations are additionally and individually organized outside of NATO, so that each of them may apply its own maritime strength in its own interests around the whole periphery of the Eurasian continent. There is difference in scale and difference in emphasis, but the underlying concepts are the same. In this struggle between East and West, the Western nations are organizing toward the full exploitation of the flexibility, resilience, endurance, and concentrated application of power that can lead to decisive control when it is needed. The whole Western world is placing its faith in the concept of a strategy that is basically maritime.

12 "BEYOND THE SEA AND JOINTNESS"

CAPT Sam J. Tangredi, USN

Awarded an Honorable Mention in the 2001 Colin L. Powell Joint Warfighting Essay Contest, this thought-provoking essay is Captain Tangredi's forward-thinking view of the post–Cold War world in which he distinguishes between the function of navies (access to the fluid mediums of a globalized world—including space and cyber—that humans use or travel through but do not inhabit) and the role of armies (control of territory). He observes that the Navy's ability to control the access to global mediums gives the Navy the ability to project power "beyond the sea." This ability is derived from being the last globally deployable navy.

No shrinking violet when it comes to controversial ideas, Captain Tangredi contends that in much of its operations (particularly space and cyber), the U.S. Air Force functions like a navy, and that the problem with jointness, as currently defined, is that roles are assigned by service, not by function. He further holds that in its breadth of operating mediums, the Navy and Marine Corps team represents *functional* jointness—beyond the current jointness concept.

"BEYOND THE SEA AND JOINTNESS"
By CAPT Sam J. Tangredi, USN, U.S. Naval Institute *Proceedings* (September 2001): 60–63.

The current concept of jointness divides the battlefield into equal air, land, and sea pieces, but in a globalized world where access is becoming more important than territory, forces that can operate across the multiple mediums of communications and exchange—the Navy, and to a lesser extent, the Air Force—will have the advantage.

In today's world, there are no navies. This might seem an overdramatic statement. There is, in fact, *a* navy. The U.S. Navy is now the only, and possibly the last, global navy. In effect, it has become the world's navy.

There are other national naval forces. The British Royal Navy and the French Navy can sustain out-of-area deployments of a battle group equivalent. Canadian, Dutch, Spanish, Italian, and Australian ships frequently conduct forward deployments, though often as adjuncts to U.S. fleet operations. Other national naval forces operate on a regional basis with the U.S. fleet. Japan has an increasingly more powerful naval force built around U.S. naval technology, although technically it is not a navy. Russia retains the hope of a navy, and China the dream of one.

But most nations neither desire nor can afford an oceangoing navy. This is a reflection of resource capabilities and the result of the Cold War victory. It provides a tremendous advantage to the security and prosperity of the United States—but it also threatens the current paradigm of jointness.

In addition, it raises a host of difficult questions. If there is but one navy, does naval warfare—traditionally defined—still exist? Have maritime operations become just a spoke in the purple umbrella of joint military operations, without their own logic and grammar? If decisive war at sea between opposing fleets is implausible, of what actual use is a navy? In the contemporary world of globalization, instant communications, and weapons of mass destruction proliferation, is there any point to discussing naval strategy at all?

To answer these questions requires a sort of deconstructionist approach. First, we must examine exactly what is a navy. Then we must look at the relationship of armies and navies in the 21st century, a period defined by the phenomenon called globalization. If, in fact, we are in a "strategic pause" that allows us the time to transform our military to face radically different future threats, then the first step is to take a radically new look at what armies and navies are designed to do. For the U.S. Navy, the conclusion is that its role no longer is tied to the physical ocean, but now lies "beyond the sea."[1] And because it lies beyond the sea, it no longer can be defined in terms of current concepts of jointness.

What Is a Navy?

The answer to this question cuts on the difference between a navy as an officially defined organization and naval operations as a military function. From the narrow perspective of organization, the obvious but only partially correct answer is that a navy is a military force that operates primarily at sea.

From this point of view, it is easy to categorize naval warfare as one corner of overall joint military operations: the army fights on land, the air force fights in the air, and the navy fights at sea. Even this simplistic formula is made more complex by the fact that U.S. naval forces also include the U.S. Marine Corps, as well as sea-based strike aviation and a host of other land-oriented functions. Yet, it is comforting for a joint planner to be able to divide the battlefield into such equal pieces. It is an image in sync with the reigning ideology of the Department of Defense: jointness defined as the relative equivalence of all missions and services.

But there is a significant difference between the functioning of navies and that of land-based military forces. Naval forces are designed primarily and uniquely to control the flow of contact through the dominant mediums of human interaction and exchange, rather than directly to control territory or areas of human habitation. In short, armies are designed to control territory; navies are designed to control access.

Fighting in a multiplicity of mediums—undersea, on the surface of the sea, in the air, the littorals, space, and the infosphere—navies contest for control of interactions rather than populations. The classical naval struggle for sea control is for dominance of oceans—not all of them wet—which are mediums that humans use but cannot permanently inhabit. Once dominated, these oceans can provide access to the areas where humans live as well as control of the links between these areas and the rest of the globe.

The difference between this concept and the organizational view is more than semantics. To occupy territory requires one to close with the enemy, defeat him, and garrison his state. Controlling access, however, involves cutting off the enemy state from the world. The fruits of being a nation-state—formalized trade relationships, interactions with other ideas and cultures, even the motivation for nationalism itself—cannot exist without interaction with the rest of the international system. These interactions require access to the fluid mediums of communications and exchange. Such access can be checked by physical blockades, interdiction, actual combat, or cyberwar or intimidated by nearby military presence.

Because of the earth's geography and the less-physical restraints of common international law, all of these things are most easily done by military forces (actually naval forces) located within those fluid mediums of exchange. To cut off access does not necessarily require closing with the enemy or occupying his territory—at least not for a sea power nation. The 17th-century political philosopher Sir Francis Bacon got it right when he said that a sea power could "take as much or as little of war as it desires."

Of course, a state of access denial in which the machinations of an aggressor are merely blocked does not appear to lead to long-term peace any more than occupation. A timely illustration is the brooding presence of Saddam Hussein, apparently unconverted by sanctions or periodic air strikes. However, this can be countered by the one huge example of an access denial strategy that worked: the containment of the Soviet Union.[2]

A navy is that portion of military forces that operates in the mediums humans use for communications, transportation, and exchange but cannot normally inhabit. Its prime purpose is to ensure or deny access. Its effect on territories and population generally is indirect; however, technology and the freedom of operation permitted in the international commons of the ocean provide for an ever-increasing reach into the littoral regions.

Under this deconstructionist definition, organizations wearing other uniforms but operating within the mediums of communications and exchange can be seen as naval in function or tone. For example, the U.S. Air Force—in its role of strategic bombing and long-range strike—operates essentially as a navy. It uses a fluid medium in which humans normally cannot survive to deny aggressors access to political goals. This is the heart of effects-based operations, even if it usually is not described in that fashion. In its interdiction and close air support roles, however, the U.S. Air Force functions like the long-range artillery of a traditional army.

This deconstructionist view of armies and navies contradicts organizational definitions and challenges some of the current jointness dogma. If armies and navies perform different functions—territory control and access control—that overlap only at the margins, can there be a joint concept of operations that integrates both equally? Or does the search for absolute jointness simply obscure the dynamic balance between territory and access control that the U.S. military had mastered during the Cold War?

Why No Other Navies?

Let's go back to the opening statement, there are no navies, and ask why.

We are defining navies as oceangoing fleets capable of sustained out-of-area power projection operations. In accordance with this approach, they also could be defined in terms of air forces capable of conducting sustained intercontinental strike or transport. As noted, most nations have given up maintaining fleets capable of sustained operations in distant regions. Similarly, intercontinental strike/bomber forces also have been reduced.

An obvious reason for the demise of navies is economics. Navies require tremendous resources to operate effectively in the fluid mediums. The main costs are maintaining the logistic capabilities required by a long-range power projection fleet, and the technology to make such a fleet combat credible. Most nations simply cannot or do not want to afford either. But perhaps even more important is the general lack of a naval threat, and hence the lack of motivation to afford an oceangoing navy. With the collapse of the Soviet Union and the fact that the United States intended to keep a superpower-sized navy, there seemed little reason for most nations to maintain an oceangoing navy at all.

The result is that the U.S. Navy can be considered a globalized as well as a global navy. In essence, it no longer is solely the United States' navy; it has become the world's navy—delivering the security of access function across the entire world system. When Asian tiger economies, such as that of Taiwan, are shaken by the bellicose posturing of a neighbor, it is the movement of U.S. naval forces into the region of potential crisis that provides the prime means of psychological restabilization.[3] Under the concept of Air Expeditionary Forces, the U.S. Air Force has moved to adopt a similar role.

In addition, with the exception of the "states formerly known as rogues" and the Chinese Communist Party, no one seems to expect any harm from the U.S. Navy. Japan, which is potentially the United States' number one economic rival, even allows the U.S. Navy to homeport both a carrier battle group and an amphibious ready group in its port cities—and pays for the infrastructure to do so. Russia, with a military still vaguely suspicious of the West, has conducted post–Cold War exercises with NATO (and U.S.) naval forces. The U.S. Navy is welcomed in ports around the globe, and the forward presence of U.S. warships is accepted—if not advocated—by most nations as a sound policy for maintaining regional security. This, again, dissuades most states from making considerable investment in navies.[4]

Extending the Effects of Access Control

Navies and armies overlap on the margins. For the U.S. Navy, this margin is growing ever wider as the reach of U.S. naval forces keeps extending over land.

The most recent evidence of this growth is the use of sea-based cruise missiles to attack and neutralize sites connected with terrorism and with the development of weapons of mass destruction. The 1998 Tomahawk strike against Bin Laden's terrorism network headquarters in Afghanistan appears the first use of naval power as the sole means of striking targets in a land-locked country. Metaphorically, this represents the projection of power from the margins of the maritime world—described as "rimlands" in earlier geopolitical theory—into the very heartland of global terrorism. Heartlands previously were the exclusive province of armies.

These developments are the tip of the iceberg of the ever-increasing ability of forces "from the sea" to direct their effect-producing efforts and energies onto land. They do not yet represent a replacement of an army's capability to occupy territory. However, this extension of access control and denial is made possible by the intertwining of three threads: an evolution in naval technology that extends the reach of naval forces, a revolution in naval affairs in which the U.S. Navy became the world's navy, and globalization.

Access as the Key to Globalization

If globalization is a "process of expansion of cross-border networks and flows," then naval forces, broadly defined, are both potential protectors and potential inhibitors of such expansion. The language of sea power—with its concern for sea lines of communications, blockades, fleets-in-being, and naval presence—may seem like a quaint legacy dialogue to those schooled in information technology and e-commerce. But though it may not use the same grammar, it uses the same logic.

The traditional goal of sea power is unfettered access to the world's common transportation routes for raw materials and manufactured products, as well as access to the actual markets and sources of materials themselves. The emerging concept of the new economy revolves around access to the world's common electronic information routes—such as the Internet—and the sources of information, as well as the potential markets for value added to the information. Like every other such shaping process, globalization, at its heart, involves a struggle for economic and political power—a struggle for access to the fruits of the process.

This struggle includes access to the infosphere, to financial markets, to raw materials (of which information is one), to the means of production, and to the market population. And just as a hacker can use information warfare to delay, disrupt, distort, or deny access to the infosphere, more traditional military forces—and primarily those that operate from within the mediums of interaction—can deny access to the sources of the production of wealth. The maintenance of a force that can operate from within the mediums, i.e., a navy, is a form of insurance that such physical access could not be cut by other military force, at least not without a war. And navies also are the means of denying access to opponents or rivals.

If globalization is breaking down the territorial barriers of our world—which is what most proponents of globalization suggest—then access to information, markets, or resources is becoming even more important to the world's political economy than control of territory, no matter how fertile or resource-filled, or control of populations, no matter how productive. This would suggest that navies are becoming more important as well. But, as was stated in the beginning, there is only one navy.

Toward a Redefinition of Jointness

Benefiting from continuing evolutions in naval technology, the U.S. Navy—as the dominant world's navy—indeed has shifted its strategic vision to using its control over access to affect events on land directly. Such a shift is made possible by the elimination of any serious challenge to U.S. sea (and air) control. This allows the U.S. Navy to focus "beyond the sea," which is a natural development because the functional definition of naval power is to be the military instrument within the fluid mediums of interaction and exchange. As these fluid mediums expand to include space and cyberspace, it is natural to use naval forces (no matter what uniform they wear) to control access.

So what does this mean for jointness?

The focus beyond the sea carries with it a shrinking of the gap between armies and navies and the potential subsuming of land power into a broader conception of sea power. This blend is made salient by the fact that in a globalized world, the United States might no longer need to control an opponent's territory to achieve its strategic effects. Thinkers within the U.S. Air Force have made similar arguments, but they have spoken solely in terms of air-centric precision strike, rather than in terms of multiple-fluid medium access control. Such a blend between land and sea (and air) power concepts would have undeniable effects on the way future joint and naval strategy are perceived. It could lead to an end to joint strategy, as currently practiced, as well as to the end of naval strategy.[5] In fact, these effects could lead to ultimate jointness, as the United States adapts its military to best use its unique advantages of access control in a globalizing world.

In the contemporary world there are no navies. That does not mean they might not be re-created in the future. There may even be some return to naval strategy as it is traditionally conceived: strategy for military forces fighting at sea. But in having to develop a naval strategy for an era without navies, the U.S. Navy is poised to take naval power beyond the constraints of time and tide and apply it, not merely from the sea, but truly beyond the sea. If it is willing to take this step, the Department of Defense may find itself in a new era of redefined jointness.

Notes

1. The phrase "beyond the sea" was the title of a proposed strategic vision, developed by members of the Strategy and Concepts Branch of the Office of the Chief of Naval Operations, that was never adopted. Credit for the phrase and many of the ideas behind it belongs to Commanders Randall G. Bowdish and Craig Faller.
2. "Access denial" is used by some as synonymous with "antiaccess" strategies, methods that regional powers might use to prevent U.S. power projection forces from entering their region. "Area denial" used here refers to the U.S. Navy's ability to cut off a potential opponent's access to the rest of the world. For discussions of antiaccess strategies, see Thomas G. Mahnken, "America's Next War," *The Washington Quarterly* 16:3 (Summer 1993): 171–84; Mahnken, "Deny U.S. Access?" U.S. Naval Institute *Proceedings* (September 1998): 36–39.
3. See Ron Brown, Robert Looney, David Schrady, et al., "Forward Engagement Requirements for U.S. Naval Forces: New Analytical Approaches," Report NPS-OR- 97–011PR (Monterey, CA: Naval Postgraduate School, July 23, 1997), and Robert Looney and David Schrady, "Estimating Economic Benefits of Naval Forward Presence" (Monterey, CA: Naval Postgraduate School, September 2000).

4. Former Secretary of the Navy Richard Danzig discusses the concept of dissuasion in *The Big Three: Our Greatest Security Risks and How to Address Them* (New York: Center for International Political Economy, February 1999), 22–24.
5. Jan Breemer has identified the revolutionary supplanting of land power by sea power as "the end of naval strategy," implying that maritime operations have become the underlying basis for all joint strategy. See Jan S. Breemer, "The End of Naval Strategy: Revolutionary Change and the Future of American Naval Power," *Strategic Review* 22:2 (Spring 1994): 40–53.

13 "WIN WITHOUT FIGHTING"

LT David A. Adams, USN

Winner of the Colin L. Powell Joint Warfighting Essay Contest, Lieutenant Adams' cogent writing in this article belies his junior rank. With a title that reflects one of the basic tenets offered by the classical strategist Sun Tzu, this article adapts that maxim to modern times and to maritime thinking. Contending that naval power is more effective than so-called "decisive force" in many real-world applications, Adams urges the adoption of a joint strategy that emphasizes "presence" and "patience."

"WIN WITHOUT FIGHTING"

By LT David A. Adams, USN, U.S. Naval Institute *Proceedings* (September 2000): 54–57.

On 6 August 1945, Colonel Paul Tibbetts Jr. piloted the B-29 bomber *Enola Gay* from the Pacific atoll of Tinian to the main island of Japan, where the bombardier dropped "Little Boy," a ten-foot-long, nine-thousand-pound device, from an altitude of thirty-two thousand feet. Exploding six hundred yards above Hiroshima, the bomb flattened four square miles of cityscape, killing more than one hundred thousand Japanese and disfiguring thousands more.[1] On 9 August, a similar device inflicted comparable carnage on Nagasaki. Our subsequent victory in the Pacific—codified by the unconditional Japanese surrender on the deck of the USS *Missouri* (BB-63)—has come to epitomize U.S. military decisiveness.

Thirty years later, on 30 April 1975, the public at home watched as U.S. helicopters frantically evacuated the last Americans to ships off the coast of South Vietnam.[2] The U.S. military also managed to extricate some six thousand South Vietnamese loyalists, but thousands more were left at the gates of the U.S. embassy, begging for sanctuary from the communist forces encircling Saigon.[3] Only hours later, Colonel Bui Tin, upon seizing Saigon, pronounced, "Between Vietnamese, there are no victors and no vanquished. Only the Americans have been beaten." The United States had indeed suffered its greatest defeat.

The experiences of World War II and Vietnam—more than any other historical events—define the American conception of military power. Conduct of the Pacific War, in particular, seemed to illustrate the use of decisive force to achieve victory. Vietnam, on the other hand, shattered our nation's postwar idealism and created doubts about the usefulness of military power. Lost in the emotional fixation on both wars has been the U.S. military's ability to formulate sound strategy. Since our withdrawal from Southeast Asia, top military leaders have demanded that U.S. forces only "apply decisive force to overwhelm our adversaries and thereby terminate conflicts swiftly with minimum loss of life."[4]

Never tested during the Cold War for fear of inciting a nuclear confrontation, the theory of decisive force has proved wholly inadequate to confront the United States' strategic challenges since. The Gulf War, for example, saw an almost-perfect application of decisive force to achieve a clearly defined objective—the liberation of Kuwait. Yet, despite an overwhelming military decision, Saddam Hussein's continued belligerence leads many Americans to wonder who actually won what. Similarly, our over reliance on "decisive" bombing and an unwillingness to risk casualties to thwart aggression on the ground in Kosovo escalated rather than deterred that humanitarian crisis, leaving five times as many homeless refugees as when NATO started bombing.[5] These phenomena reinforce what strategists long have known: victory does not lie in the realm of decisive force, but in the more perplexing world of politics. The demands of a new century require us to move beyond decisive force toward a more sophisticated strategy.

Joint Vision or Strategic Blindness?

Unfortunately, U.S. strategic vision is clouded not only by memories of past wars but also by the current fixation on jointness. Some military officers have pointed out that the services' war fighting tactics differ greatly,[6] but they miss a larger point: the political dynamics of pursuing naval (maritime), military (continental), and air strategies have even less in common. Each service cultivates a very different conception of how to derive political benefit from the threat or use of force. Attempts to create a monolithic joint culture have denied the country an array of strategic options that naturally flow from the unique service heritages.

Goldwater-Nichols may have tempered inter-service rivalry but it also unwittingly silenced the strategic debate. Doctrinal power today is centered with the Chairman of the Joint Chiefs of Staff, a position held by Army generals for more than a decade now. These Army leaders—who have had difficulty in getting past their experiences in Vietnam[7]—have cemented decisive force into both joint doctrine and the National Military Strategy. Air Force planners, still enamored of the possibility of decisive bombing, have been more than willing to go along. Surprisingly, even Navy leaders have bought off on decisive force, in spite of the fact that it runs completely contrary to a more appropriate naval strategy.[8]

Decisive force has little significance at a time when the United States faces no great continental adversaries. World instabilities—not great power confrontations—are the greatest threats to peace and prosperity today.[9] U.S. foreign and military policies no longer need to focus primarily on war but instead on a classic maritime problem—how to project U.S. influence to promote stability, prevent conflicts, and foster a favorable climate for our expanding markets. In these endeavors, Clausewitz's *On War* is strikingly empty of practical guidance.[10]

By contrast, when confronted with the limitations of sea power, famous naval thinkers became the true masters of strategy. *The Influence of Sea Power on History*, for example, offers a coherent explanation of how an inherently limited naval force can yield lasting economic and political benefits for a nation.[11] Much of this argument still applies in a joint context. Thus, given today's strategic circumstances, all military strategists would do better to listen to Mahan than Clausewitz. The 21st century calls for a naval approach to joint U.S. strategy.

Rethinking Past Conflicts

History has shown that a committed presence is key to influencing international outcomes. A prolonged commitment to overseas presence after World War II and during the Cold War, for example, produced some of our most profound international successes. In Vietnam, on the other hand, our quest for decisive victory in the name of "winning quickly and bringing our boys home" guaranteed failure by assuring the enemy that our presence was fragile and temporary. A closer examination of these wars reveals that the outcomes had little to do with the prevalence or paucity of decisive blows.

Victory in the Pacific. The unconditional surrender of the Japanese while they still possessed a viable two-million-man army was unprecedented in the history of conflict. Unfortunately, the timing of the surrender, immediately after the dropping of the atomic bombs on Nagasaki and Hiroshima, will forever muddle our perception of the event. Imbedded in U.S. strategic culture is the fable that the atomic bombs were decisive and the Japanese surrendered to avoid nuclear obliteration. These popular acumen are not supported by the facts.[12]

Consider that the United States already had flattened virtually every major Japanese city. Almost two hundred square miles of Japan had been relentlessly fire bombed prior to the atomic attacks. These conventional attacks left a third of all Japanese homeless and resulted in seven times the fatalities of Hiroshima and Nagasaki. Through all this destruction the civilian populous remained passive and Japanese leaders failed to urgently pursue surrender. Additional atomic attacks could not possibly have inflicted significantly more damage; in effect "the hostage was already dead." This helps to explain why Japan's Supreme War Council did not even meet after the nuclear attack on Hiroshima.[13]

By contrast, the Soviet declaration of war on 8 August produced an immediate War Council meeting and the subsequent decision to surrender.[14] From the start the Japanese plan for victory had been to strike, seize, secure, and negotiate: strike Pearl Harbor, destroy the U.S. Pacific Fleet, and expel the Americans from the Philippines; seize much of Southeast Asia before the United States could recover and leverage those newly acquired resources to end the stalemate in China; secure control of the Pacific; and negotiate a peace using the Soviets as intermediary.

By the time the atomic bombs were dropped, this strategy had been almost completely repudiated. Having awakened a sleeping industrial giant, Japan had been pushed out of Southeast Asia, was bogged down in a dirty war in China, and no longer had the industrial capacity—thanks in large measure to commerce raiding by U.S. submarines—to defend the home islands. Clay Blair was right, "the atomic bombs were simply the funeral pyre of an enemy that had already been sunk."[15]

Realizing the war was lost, the emperor was desperately seeking Soviet help to end the conflict in the summer of 1945. Instead, the Soviets declared war. This both ensured the eventual defeat of the Japanese military and denied the Imperial court its planned intermediary. In the end, the Japanese surrender was less the result of a decisive atomic blow than a culmination of factors that left the Japanese without hope of bettering the final outcome by continuing to fight.

Considering our military triumph, it is easy to overlook that our greatest victory was not on the battlefield. Most decisive was our commitment to the MacArthur Plan, which allowed the United States to reap long-term political benefits from our hard-won battles. A prolonged U.S. presence in Japan was key to creating a stable balance of power in Asia and to encouraging Japan and other Asian nations to move toward market economies and democratic governments. Maritime encirclement and prolonged presence—key elements of naval strategy—ensured a favorable and lasting resolution to our conflict with Japan.

Defeat in Vietnam. The causes of U.S. failure in Vietnam also are widely misunderstood. The most common explanation for the loss is the half-truth that the "politicians tied the military's hands," and some military experts still believe that to win we needed only to unleash our overwhelming force against the North Vietnamese.[16] It is easy to forget, however, that the French occupied most of North Vietnam and still suffered a humiliating defeat.[17] Lacking a more sophisticated strategy, no amount of decisive force could have produced success in Vietnam. In fact, the U.S. military's obsession with decisive victory ultimately was its undoing.

Early in the war both our political and our military leaders lost sight of the nation's singular strategic objective in Vietnam: to preserve a noncommunist government in Saigon. This objective was consistent with our national strategy of containment and may have been achievable given strong military and political leadership. Containment was, by its very nature, defensive. It required us only to hold the line against communism—not to annihilate the insurgents in the South or to defeat their communist sponsors in the North.

Any credible long-term commitment of U.S. economic and military resources to Vietnam would have ensured attainment of our primary objective. It is inconceivable that a communist takeover could have occurred with any substantial U.S. forces in South Vietnam, and the North's numerous failed attempts to incite a massive popular revolt in the South demonstrated that the communist insurgency had failed in the light of U.S. presence.[18] But instead of emphasizing presence and limited force to preserve the South Vietnamese government, the U.S. military turned to large-scale conventional operations in an attempt to eradicate the enemy. Lack of a clear policy combined with the needless escalation of casualties turned widespread public support for the war to skepticism.

Prominent officers exacerbated the social crisis at home by telling the public that "the enemy no longer has the ability to mount an offensive."[19] Then came the Tet Offensive. Although it was a military disaster for the communists in-country, Tet had the powerful side effect of convincing Americans that their leaders were either dishonest or inept. The communists were far from finished, and exposure of the lie that the war could be won quickly made defining coherent goals in Vietnam impossible. Popular support eroded. In the end, the United States seized a dishonest peace, deserted the government it had created, and left thousands of Vietnamese allies to face communist tyranny alone. By focusing on the "need" for decisive victory rather than on the more achievable goal of maintaining a noncommunist regime, our leaders played into the enemy's hands and led the nation down a path to bitter defeat.[20]

The Myth of Decisive Force

Defeat in Vietnam left military professionals disillusioned and longing for the decisiveness of World War II. Many future military leaders turned to Clausewitz for answers. For all his tactical brilliance, however, Clausewitz did not have a sophisticated view of the linkage between the operational and political realms of conflict. He was at a loss to explain, for example, why despite Napoleon's battlefield decisiveness, the French Emperor repeatedly failed to reap lasting political advantage from his widespread military victories. Clausewitz's theory of decisive battle nevertheless validated air- and land-minded services' preconceived notions about the nature of warfare.[21] Vietnam era reformers rose in rank and established decisive force as the new American way of war.[22]

Pursuing decisive force does not make for sound strategy. Nevertheless, the concept is appealing because it offers false hope that military action can produce quick results with little bloodshed. The problem is that military force, by itself, rarely is decisive in a political sense. Clausewitz recognized that war ultimately is political, but he failed to emphasize that while military engagements might sometimes be won quickly, political attitudes are always slow to change. In this light, any theory that promises quick and lasting results at little cost in terms of blood, toil, and treasure is bound to miss the mark. In formulating strategy, there is no substitute for an appreciation of time, tenacity, and a prolonged national commitment that often must continue long after the fighting stops.

Toward a Naval Approach

Naval theory, exemplified by the works of Alfred Thayer Mahan, embodies these characteristics. Mahan's writings are not without tactical error—critics are too quick to point out his neglect of direct power projection ashore and submarine *guerre de course*, for example[23]—but *The Influence of Sea Power* was a triumph of strategic thinking. His tactical lapses do not diminish a brilliant explanation of how military power can influence international discourse.

Naval strategies in the Mahanian tradition have been so successful because they enlist time as an ally in pressuring other nations through prolonged presence, maritime encirclement, and the constant threat of power projection. It is no coincidence that great sea powers or maritime coalitions have won every major war in modern history.[24] Unlike their air- and land-minded brethren, navalists always have understood that good strategy is never quick but must seek to influence events over time.[25]

Strategic success almost always is a function of sustained presence and combined arms. Consider that the United States won the Cold War using essentially a naval approach to joint strategy. Containment was a "blockade" of communist expansion accomplished by joint forces. Committed presence, patience, and the operational ability to instantly project power resulted in tremendous economic and military pressure that eventually took their toll. We achieved what Sun Tzu coined the acme of strategy, "to win without fighting."[26]

Mahanian principles, then, offer solid underpinning for an effective joint strategy. The challenge is to convince all the services to dispel the myth of decisive force in favor of this new approach. That may not be easy.

The Problems with Decisive Force

- Lays the onus on politicians to establish clearly defined objectives, thereby allowing U.S. military leaders to avoid the question of how to use force for political gain.

- Provides cover for military leaders who would rather not answer when the political ramifications of using force do not work out quite as expected.

- Drives the U.S. military to focus on large-scale conventional conflicts at the expense of developing strategies aimed at succeeding in the more likely limited conflicts.

- Limits the military options presented to policymakers; innovative joint operations can achieve strategic effect far beyond that of bombing alone without committing to full-scale occupation.

- Undermines the prolonged U.S. presence that has proved essential to real victories over the long haul.

Decisive force advocates generally have held presence and influence in disdain, but underestimating their importance is a great strategic mistake.[27] History has shown that a credible U.S. presence is needed in key regions to guarantee peace and stability, to keep the nation secure, and to allow our markets to flourish. It is time for officers of all services to embrace the naval approach because it offers the best hope to shape political attitudes over time, to deter conflicts, and to limit violence if political differences turn hostile. The alternative is reacting to regional disturbances with inappropriate explosions of "decisive" force that leave political differences unresolved and festering. The consequence could be a new century even more bloody and chaotic than the last.

Notes

1. James Trager, *The People's Chronology* (New York: Henry Holt, 1994), 893.
2. Robert D. Schulzinger, *A Time for War: The United States and Vietnam, 1941–1976* (New York: Oxford University Press, 1997), 327.
3. Trager, *The People's Chronology*, 1049.
4. David J. Andre, "National Culture and Warfare—Wither Decisive Force," *Joint Forces Quarterly* (Autumn 1996): 105. This is Andre's synopsis of the underlying concept of the 1992 National Military Strategy of the United States.
5. Michael Evans, "Dark Victory," U.S. Naval Institute *Proceedings* (September 1999): 33–37.
6. Cdr. Terry C. Pierce, USN, "Teaching Elephants to Swim," U.S. Naval Institute *Proceedings* (May 1998): 26–29.
7. A point made nicely by Colin S. Gray, *Explorations in Strategy* (London: Greenwood Press, 1999), 236.
8. Col. Mark Cancian, USMCR, "Centers of Gravity Are a Myth," U.S. Naval Institute *Proceedings* (September 1998): 33. Cancian explains that classic naval theory is not compatible with "new joint theories of centers of gravity, direct attack, or quick decisions."
9. See, for instance, Samuel P. Huntington, *The Clash of Civilizations* (New York: A Touchstone Book, 1997), and Robert D. Kaplan, "The Coming Anarchy," *Atlantic Monthly* 273 (February 1994): 44–76.
10. Carl Von Clausewitz, *On War* (New York: Penguin Books, 1968). Only a few have pointed out the strategic inconsistencies of his works. See B. H. Liddell Hart, *Strategy* (New York: A Meridian Book, 1991), 208–212, for a solid critique.

11. Alfred Thayer Mahan, *The Influence of Sea Power on History, 1600–1783* (Boston: Little, Brown, 1918).
12. Robert A. Pape, "Why Japan Surrendered," *International Security* 18, no. 2: 154. Pape is convincing that the atomic bombs were not the decisive factor. The argument that follows is a synopsis of this fine work.
13. Pape, "Why Japan Surrendered," 165, 155.
14. Ibid., 155.
15. Clay Blair, *Silent Victory: The U.S. Submarine War against Japan* (New York: Bantam Books, 1975), 19.
16. See Harry G. Summers Jr., *On Strategy* (New York: Dell, 1982).
17. Schulzinger, *A Time for War*, 58–68.
18. Stanley Karnow, *Vietnam: A History* (New York: Penguin Books, 1984), 538–558.
19. Karnow, *Vietnam*, 556.
20. NA (note refers to caption that appeared in original article but has not been included here).
21. Carl Builder, *The Masks of War* (Baltimore: The Johns Hopkins University Press, 1989), 57–73.
22. For a complete analysis see F. G. Hoffman, *Decisive Force: The New American Way of War* (Westport, CT: Praeger Press, 1996).
23. See, for example, Philip A. Crowl, "Alfred Thayer Mahan: The Naval Historian," in *The Makers of Modern Strategy* (Princeton, NJ: Princeton University Press, 1986), 444–477.
24. This is the well-supported thesis of Colin S. Gray, *The Leverage of Sea Power* (New York: The Free Press, 1992).
25. Cancian, "Centers of Gravity Are a Myth," 32.
26. Sun Tzu, *The Art of War*, trans. Samuel B. Griffith (London: Oxford University Press, 1963), 75.
27. General Powell's affinity for decisive force is typical of Army officers of his generation. His quote is from Colin Powell, *My American Journey* (New York: Random House, 1995), 147.

14 "NOTES ON STRATEGY"

(Selection from the Appendix to the Classics of Sea Power edition of *Some Principles of Maritime Strategy*)

Sir Julian Corbett

Sir Julian Corbett shares top honors in classical maritime strategy writing with Alfred Thayer Mahan. Unlike Mahan, Corbett was not a naval officer (he was a lawyer) and not American (he was British) but, like Mahan, he is frequently considered the foremost naval thinker by his nation. His *Some Principles of Maritime Strategy* is considered by many to be superior to Mahan's writings and is a staple of the U.S. Naval War College's "Strategy and Policy" curriculum. Although considerably shorter than Mahan's work, it is too long for inclusion in this collection, but the Naval Institute's Classics of Sea Power version (which was introduced by an informative essay by renowned scholar Eric Grove) included as an appendix a student handout that Corbett used while teaching at the Royal Naval War College that is both concisely informative of Corbett's thinking and brief enough to be included in this anthology. Known in its day and ever since as "The Green Pamphlet," these notes were certainly precursors to the wisdom contained in Corbett's seminal work.

"NOTES ON STRATEGY"

(Selection from the Appendix to the Classics of Sea Power edition of *Some Principles of Maritime Strategy*) by Sir Julian Corbett (Naval Institute Press, 1988): 326–45.

PART ONE
General Principles and Definitions

Introductory

Naval strategy is a section of the Art of War.

The study for officers is the Art of War, which includes Naval Strategy.

War is the application of force to the attainment of political ends.

Major and Minor Strategy

We seek our ends by directing force upon certain objects, which may be ulterior or primary.

Primary objects are the special objects of particular operations or movements which we undertake in order to gain the ulterior object of the campaign. Consequently it must be remembered that every particular operation or movement must be regarded, not only from the point of view of its special object, but also as a step to the end of the campaign or war.

Strategy is the art of directing force to the ends in view. There are two kinds—Major Strategy, dealing with ulterior objects; Minor Strategy, with primary objects.

Every operation of an army or fleet must be planned and conducted in relation (1) to the general plan of the war; (2) to the object to which it is immediately directed.

Major Strategy, always regarding the ulterior object, has for its province *the plan of the war* and includes: (1) Selection of the immediate or primary objects to be aimed at for attaining the ulterior object; (2) Selection of the force to be used, *i.e.,* it determines the relative functions of the naval and military forces.

Major Strategy in its broadest sense deals with the whole resources of the nation for war. It is a branch of statesmanship which regards the Army and Navy as parts of one force, to be handled together as the instrument of war. But it also has to keep in constant touch with the political and diplomatic position of the country (on which depends the effective action of the instrument), and the commercial and financial position (by which the energy for working the instrument is maintained). The friction due to these considerations is inherent in war, and is called the deflection of strategy by politics. It is usually regarded as a disease. It is really a vital factor in every strategical problem. It may be taken as a general rule that no question of major strategy can be decided apart from diplomacy, and *vice versâ*. For a line of action or an object which is expedient from the point of view of strategy may be barred by diplomatic considerations, and *vice versâ*. To decide a question of Major Strategy, without consideration of its diplomatic aspect, is to decide on half the factors only. Neither strategy or diplomacy has ever a clean slate. This inter-action has to be accepted as part of the inevitable "friction of war." A good example is Pitt's refusal to send a fleet into the Baltic to assist Frederick the Great during the Seven Years' War, for fear of compromising our relations with the Scandinavian Powers.

Minor Strategy has for its province *the plans of operations*. It deals with—

(1) The selection of the "objectives," that is, the particular forces of the enemy or the strategical points to be dealt with in order to secure the object of the particular operation.
(2) The direction of the force assigned for the operation.

Minor Strategy may, therefore, be of three kinds:—

(1) Naval, where the immediate object is to be attained by a fleet only.
(2) Military, where the immediate object is to be attained by an army only.
(3) Combined, where the immediate object is to be attained by army and navy together.

It will be seen, therefore, that what is usually called Naval Strategy or Fleet Strategy is only a sub-division of Strategy, and that therefore Strategy cannot be studied from the point of view of naval operations only. Naval Strategy, being a part of General Strategy, is subject to the same friction as Major Strategy, though in a less degree. Individual commanders have often to take a decision independently of the central government or headquarters; they should, therefore, always keep in mind the possible ulterior effects of any line of action they may take, endeavouring to be sure that what is strategically expedient is not diplomatically inexpedient.

Example.—For example, take Boscawen's attack on De la Motte on the eve of the Seven Years' War in 1755. His orders were to prevent the troops and warlike stores which De la Motte was taking out from reaching Canada. It was not diplomatically expedient to open hostilities; but if Boscawen succeeded, the result would have been worth the diplomatic consequences it would entail. He missed the expedition, but captured two isolated vessels; thus striking the first blow in such a way as to entail the utmost amount of harm with the least possible good.

Offensive and Defensive
Nature of Object

Upon the nature of the object depends the fundamental distinction between *offensive* and *defensive*, upon which all strategical calculation must be based. Consequently, the solution of every strategical problem, whether of Major or Minor Strategy, depends primarily on the nature of the object in view.

All objects, whether ulterior or not, may be positive or negative.

A *positive object* is where we seek to assert or acquire something for ourselves.

A *negative object* is where we seek to deny the enemy something or prevent his gaining something.

Where the object is positive, Strategy is offensive.

Where the object is negative, Strategy is defensive.

This is the only certain test by which we can decide whether any particular operation is offensive or defensive.

Ulterior objects are not necessarily the same in their nature as the primary or secondary objects which lead up to them; *e.g.,* ulterior objects may be offensive, while one or more of the primary objects may be defensive, and *vice versâ.* For example, in the Russo-Japanese War the ulterior object of the war (to drive Russians from Manchuria) was offensive (positive). The ulterior object of the fleet (to cover the invasion) was defensive (negative). Its primary object to effect this was to attack and destroy the Russian naval force. This was offensive (positive).

Relation of Offensive to Defensive

The Offensive, being positive in its aim, is naturally the more effective form of war and, as a rule, should be adopted by the stronger Power.

The Defensive, being negative in its aim, is the more lasting form of war, since it requires less force to keep what one has than to take what is another's, and, as a rule, is adopted by the weaker Power.

In most cases where the weaker side successfully assumes the offensive, it is due to his doing so before the enemy's mobilization or concentration is complete, whereby the attacking force is able to deal in succession with locally inferior forces of the enemy.

The advantages of the Offensive are well known.

Its disadvantages are:—

> That it grows weaker as it advances, by prolonging its communications, and that it tends to operations on unfamiliar ground.

The advantages of the Defensive are chiefly:—

> Proximity to the base of supply and repair stations, familiar ground, facility for arranging surprise by counter attack, and power of organising in advance.

The disadvantages of the Defensive are mainly moral. They become, however, real and practical when the enemy's objective or line of operations cannot be ascertained, for then we have to spread or attenuate our force to cover all probable objectives, but this disadvantage can be neutralised when it is possible to secure an interior position.

Functions and Characteristics of the Defensive

True Defensive means waiting for a chance to strike.

To assume the defensive does not necessarily mean that we do not feel strong enough to attack. It may mean that we see our way by using the defensive to force certain movements on the enemy which will enable us to hit harder.

A well-designed defensive will always threaten or conceal an attack. Unless it does this it will not deflect the enemy's strategy in our favour. Thus, in 1756, the French, by assuming the defensive in the Channel, threatened an attack on our coasts, and concealed their attack on Minorca.

This power inherent in the defensive is peculiarly strong in naval warfare, since the mobility of fleets enables them to pass instantaneously from the defensive to the offensive without any warning.

When we assume the defensive because we are too weak for the offensive, we still do not lay aside attack. The whole strength and essence of the defensive is the counter-stroke. Its cardinal idea is to force the enemy to attack us in a position where he will expose himself to a counter-stroke.

The stock instance upon which naval defensive is usually condemned is the burning of our ships at Chatham by the Dutch. But in that case we were not *acting on the defensive* at all. We had laid up our battle fleet and were doing nothing. We were purely passive, in expectation of peace. It is really an instance of the successful use of defensive *by the Dutch*. Being no longer strong enough for a general offensive, they assumed the defensive, and induced us to lay up our ships and so expose ourselves to a counter-stroke. It was a counter-stroke by the worsted belligerent to get better terms of peace.

So far is the defensive from excluding the idea of attack, that it may consist entirely of a series of minor offensive operations. Clausewitz calls it "a shield of blows." It is often called *offensive-defensive,* or *active defence.* Neither term is really necessary. For a defensive which excludes the idea of offence or action is not war at all—at least at sea. The old Elizabethan term *Preventive* most closely expresses the idea.

The most important function of the defensive is that of covering, buttressing, and intensifying the main attack. No plan of campaign, however strong the offensive intention, is perfect which does not contemplate the use of the defensive. Without some use of the defensive the cardinal principle of concentration can rarely be fully developed. To develop the highest possible degree of concentration upon the main object or objective, the defensive must be assumed everywhere else. Because it is only by using the defensive in the minor or less important theatres of operation that the forces in those theatres can be reduced to the minimum of security, and the maximum of concentration can thereby be obtained in the main theatre.

In considering the defensive as a general plan of campaign the maxim is: If not relatively strong enough to assume the offensive, assume the defensive till you become so—

(1) Either by inducing the enemy to weaken himself by attacks or otherwise;
(2) Or by increasing your own strength, by developing new forces or securing allies.

It must always be remembered that, except as a preparation or a cover for offensive action, the defensive is seldom or never of any use; for by the continued use of the defensive alone nothing can be acquired, though the enemy may be prevented from acquiring anything. But where we are too weak to assume the offensive it is often necessary to assume the defensive, and wait in expectation of time turning the scale in our favour and permitting us to accumulate strength relatively greater than the enemy's; we then pass to the offensive, for which our defensive has been a preparation.

At sea we have had little occasion for the defensive as a general plan. But that is no reason for neglecting its study. In despising the defensive ourselves we have consistently ignored the strength it gives our enemies. The bulk of our naval history is the story of how we have been baffled and thwarted by our enemies assuming the defensive at sea in support of their offensive on land. We have seldom succeeded in treating this attitude with success, and it is only by studying the defensive we can hope to do so.

Offensive Operations Used with a Defensive Intention

(A) Counter attacks.
(B) Diversions.

Counter attacks are those which are made upon an enemy who exposes himself anywhere in the theatre of his offensive operations. It is this form of attack which constitutes what Clausewitz calls the "surprise advantage of defence."

Diversions are similar operations undertaken against an enemy outside the limit of his theatre of offensive operations.

Diversions are designed to confuse his strategy, to distract his attention, and to draw off his forces from his main attack. If well planned, they should divert a force greater than their own. They should, therefore, be small. The nearer they approach the importance of a real attack the less likely they are to divert a force greater than their own.

Diversions involve a breach of the law of concentration, and it is only their power of diverting or containing a larger force than their own that justifies their use.

This power depends mainly on suddenness and mobility, and these qualities are most highly developed in combined expeditions.

Diversions must be carefully distinguished from *eccentric attacks*. Eccentric attacks are true offensive movements. They have a positive object, *i.e.*, they aim to acquire something from the enemy; whereas diversions have a negative object, *i.e.*, they aim at preventing the enemy doing or acquiring something. Eccentric attacks are usually made in greater force than diversions.

Examples.—*Diversion.*—Our raid on Washington in 1815. Landing force, about 4,000 men. Object, according to official instruction, a "diversion on the coasts of United States of America in favour of the army employed in the defence of Canada"; *i.e.*, the intention was negative—preventive—defensive. *Eccentric Attack.*—Operations against New Orleans in 1815. Intended force, 15,000 to 20,000 men. Object, "to obtain command of the embouchure of the Mississippi, and, secondly, to occupy some important and valuable possession, by the restoration of which the conditions of peace might be improved, &c."; *i.e.*, the intention was positive—to acquire. Compare Rochefort Expedition (diversion) with those against Martinique and Belleisle (eccentric attacks) in the Seven Years' War.

This distinction gives a threefold classification of combined expeditions, as used by Elizabethan strategists, viz., raids, incursions, and invasions. These correspond respectively with our modern diversions, eccentric attacks, and true direct offensive.

Limited and Unlimited Wars

From the nature of the ulterior object we get an important classification of wars, according to whether such object is *limited* or *unlimited.*

(1) *War with limited object* ("limited war") is where the object is merely to take from the enemy some particular part of his possessions or interests; *e.g.,* Spanish-American War, where the object was the liberation of Cuba.

(2) *War with an unlimited object* is where the object is to overthrow the enemy completely, so that to save himself from destruction he must agree to do our will (become subservient); *e.g.,* Franco-German War.

Plans of War

System of Operations

Having determined the nature of the war by the nature of its object (*i.e.,* whether it is offensive or defensive and whether it is limited or unlimited), Strategy has to decide on the system of operations or "plan of the war."

Apart from the means at our disposal a plan of war depends mainly upon—

(1) The theatre of the war.
(2) The various theatres of operation available within it.

1. *Theatre of the War.*—Usually defined as "All the territory upon which the hostile parties may assail each other." This is insufficient. For an Island Power the theatre of war will always include sea areas. Truer definition: "Geographical areas within which must lie the operations necessary for the attainment of the ulterior objects of the war and of the subordinate objects that lead up to them."

A "theatre of war" may contain several "theatres of operations."

2. *Theatre of Operations.*—Is generally used of the operations of one belligerent only.

An "operation" is any considerable strategical undertaking.

A "theatre of operations" is usually defined as embracing all the territory we seek to take possession of or to defend.

A truer definition is: "The area, whether of sea or land or both, within which the enemy must be overpowered before we can secure the object of the particular operation."

Consequently, since the nature of the war varies with the object, it may be defensive in one theatre of operations and offensive in another.

Where the operations are defensive in character any special movement or movements may be offensive.

As the plan of war determines the theatres of operation in the theatre of war, so in each theatre of operation it determines the *lines of operation* and the *objectives*.

Objective

An objective is "any point or force against which an offensive movement is directed." Thus, where the *object* in any theatre of operation is to get command of a certain sea in which the enemy maintains a fleet, that fleet will usually be the *objective*.

"Objective" in ordinary use is frequently confused with "object." For purposes of strategical discussion it is desirable to keep them sharply distinguished. *Objective* is the end of some particular movement or operation, and is the special concern of the officer in command. *Object* is the end of a system of operations and is the special concern of the general staff or director of the war. An *objective* is some definite point which we wish to get from the enemy or prevent his occupying, or some part of his strength which we wish to destroy. It is incorrect to use the term of anything we already possess. Thus, Vladivostock is often said to have been Rojesvensky's *objective*. It was, strictly speaking, only his *destination*. To reach it and concentrate with the units already there was the *primary object* of the operations entrusted to him. He had no true *objective* before him except Togo's fleet.

An *objective* is always subordinate to some *object*. It is a step to the attainment of that object.

Lines of Operation

A line of operation is "the area of land or sea through which we operate from our base or starting point to reach our objectives."

Lines of operation may be *exterior* or *interior*. We are said to hold the *interior lines* when we hold such a position, in regard to a theatre of operations, that we can reach its chief objective points, or forces, more quickly than the enemy can move to their defence or assistance. Such a position is called an *interior position*. "Exterior Lines" and "Exterior Positions" are the converse of these.

Lines of Communication

This expression is used of three different things:—

(1) *Lines of supply*, running from the base of operations to the point which the operating force has reached.
(2) *Lines of lateral communication* by which several forces engaged in one theatre of operations can communicate with each other and move to each other's support.
(3) *Lines of retreat*, which are usually lines of supply reversed, *i.e.*, leading back to the base.

For naval purposes these three ideas are best described by the term "lines of passage and communication," which were in use at the end of the 18th century, and they may be regarded as those waters over which passes the normal course of vessels proceeding from the base to the objective or the force to be supplied.

Maritime Communications

The various kinds of Maritime Communications for or against which a fleet may have to operate are:—

(1) Its own communications, or those of its adversary (which correspond to the communications of armies operating ashore). These have greatly increased in importance strategically with the increased dependence of modern fleets on a regular supply of coals, stores, ammunition, &c.
(2) The communications of an army operating from an advanced oversea base, that is, communication between the advanced and the main base.

(3) Trade Routes, that is, the communications upon which depend the national resources and the supply of the main bases, as well as the "lateral" or connecting communications between various parts of belligerents' possessions.

In Land Strategy the great majority of problems are problems of communication. Maritime Strategy has never been regarded as hinging on communications, but probably it does so, as will appear from a consideration of Maritime Communications, and the extent to which they are the main preoccupation of naval operations; that is to say, all problems of Naval Strategy can be reduced to terms of "passage and communication," and this is probably the best method of solving them.

PART TWO

Naval Strategy Considered as a Question of Passage and Communication

Naval Strategy Defined

By "Naval Strategy" we mean the art of conducting the major operations of the fleet. Such operations have for their object "passage and communication"; that is, the fleet is mainly occupied in guarding our own communications and seizing those of the enemy.

We say the aim of Naval Strategy is to get command of the sea. This means something quite different from the military idea of occupying territory, for the sea cannot be the subject of political dominion or ownership. We cannot subsist upon it (like an army on conquered territory), nor can we exclude neutrals from it. The value of the sea in the political system of the world is as a means of communication between States and parts of States. Therefore the "command of the sea" means the control of communications in which the belligerents are adversely concerned. The command of the sea can never be, like the conquest of territory, the ulterior object of a war, unless it be a purely maritime war, as were approximately our wars with the Dutch in the 17th century, but it may be a primary or immediate object, and even the ulterior object of particular operations.

History shows that the actual functions of the fleet (except in purely maritime wars) have been threefold:—

1. The furtherance or hindrance of military operations ashore.
2. The protection or destruction of commerce.
3. The prevention or securing of alliances (*i.e.,* deterring or persuading neutrals as to participating in the war).

Examples.—The operations of Rooke in the first years of the War of the Spanish Succession, 1702–04, to secure the adhesion of Savoy and Portugal to the Grand Alliance. Operations of Nelson to maintain the alliance of the Kingdom of Naples.

In the first case, there came a crisis when it was more important to demonstrate to Savoy and Portugal what they stood to lose by joining Louis XIV, than to act immediately against the Toulon Fleet. In the second, the Neapolitan Alliance was essential to our operations in the Eastern Mediterranean; the destruction of the Toulon Fleet was not.

In this way we get a *Definition of the Aim of Naval Strategy,* expressed in terms of the actual functions of the fleet. For practical purposes it will be found the most useful definition as emphasising the intimate connection of Naval Strategy with the diplomatic, financial, and military aspects of Major Strategy.

These functions of the fleet may be discharged in two ways:—

(1) By direct territorial attacks, threatened or performed (bombardment, landings, raiding parties, &c.).
(2) By getting command of the sea, *i.e.,* establishing ourselves in such a position that we can control the maritime communications of all parties concerned, so that we can operate by sea against the enemy's territory, commerce, and allies, and they cannot operate against ours.

The power of the second method, by controlling communications, is out of all proportion to that of the first—direct attack. Indeed, the first can seldom be performed with any serious effect without the second. Thus, from this point of view also, it is clear that Naval Strategy is mainly a question of communications.

But not entirely. Circumstances have arisen when the fleet must discharge part of its function by direct action against territory before there is time to get general control of the communications. (That is, political and military considerations may deflect the normal operation of Naval Strategy.)

Examples.—Rooke's capture of Gibraltar in 1704, in the face of the unshaken Toulon Fleet. Holmes's capture of Emden in 1758.

Still, the fact remains that the key to the effective performance of the fleet's duties is almost always to secure communications as soon as possible by battle.

Command of the Sea

Command of the sea exists only in a state of war. If we say we have command of the sea in time of peace it is a rhetorical expression meaning that we have adequate naval positions, and an adequate fleet to secure the command when war breaks out.

Command of the sea does not mean that the enemy can do absolutely nothing, but that he cannot *seriously* interfere with the undertakings by which we seek to secure the object of the war and to force our will upon him.

Various Conditions of Command

1. It may be (*a*) general; (*b*) local.
    (a) *General command* is secured when the enemy is no longer able to act dangerously against our line of passage and communication or to use or defend his own, or (in other words) when he is no longer able to interfere seriously with our trade or our military or diplomatic operations. This condition exists practically when the enemy is no longer able to send squadrons to sea.
    (b) *Local command* implies a state of things in which we are able to prevent the enemy from interfering with our passage and communication in one or more theatres of operation.

2. Both general and local command may be (*a*) temporary; (*b*) permanent.

(a) *Temporary command* is when we are able to prevent the enemy from interfering with our passage and communication in all or some theatres of operation during the period required for gaining the object in view (*i.e.*, the object of a particular operation or of a particular campaign). This condition existed after Togo's first action. It was also that at which Napoleon aimed in his instructions to Villeneuve in 1805.

(b) *Permanent command* is when time ceases to be a vital factor in the situation, *i.e.*, when the possibility of the enemy's recovering his maritime position is too remote to be a practical consideration. This condition existed after Tsushima.

Command in Dispute

The state of dispute is the most important for practical strategy, since it is the normal condition, at least in the early stages of the war, and frequently all through it.

The state of dispute continues till a final decision is obtained, *i.e.*, till one side is no longer able to send a squadron to sea.

It is to the advantage of the preponderating Navy to end the state of dispute by seeking a decision. Hence the French tradition to avoid decisive actions as a rule when at war with England.

It must be remembered that *general command of the sea is not essential to all oversea operations.*

In a state of dispute the preponderating Power may concentrate or be induced to concentrate in one theatre of operations, and so secure the local or temporary command sufficient for obtaining the special object in view, while the weaker Power takes advantage of such local concentration to operate safely elsewhere.

Thus in a state of dispute, although the weaker Power may not be able to obstruct the passage and communication of the stronger, it may be able to defend its own.

Examples.—This condition of dispute existed during the first three years of the Seven Years' War, until Hawke and Boscawen obtained a decision by defeating Conflans and De la Clue; also in the Great War up to Trafalgar.

When the preponderating Power fails or neglects to get command (*i.e.*, leaves the general command in dispute), the disadvantage to him is not so much the danger to his own operations as the facility given to the enemy for carrying out counter operations elsewhere.

Methods of Securing Control

1. *Permanent general control* can only be secured by the practical annihilation of the enemy's fleet by successful actions.
2. *Local and temporary control* may be secured by—
    (*a*) An action not necessarily entirely successful (containing).
    (*b*) Inducing concentration on the enemy elsewhere (diversion).
    (*c*) Superior concentration so as to render impotent the enemy's force available in the special theatre of operations (masking or containing).
    (*d*) Blockade.

Action of a Fleet off an Enemy's Port

A belligerent fleet off an enemy's port may carry out three different operations, for certain purposes; each quite separate from the others, and intended to obtain an entirely different result:—

(1) *Close Blockade.*—This is to prevent the enemy's fighting ships from putting to sea. In this case the object is to secure local control for some purpose that is not purely naval, such as was carried out by the Japanese off Port Arthur in 1904, so as to enable their transports to cross the Yellow Sea without fear of molestation from any of the Russian ships in Port Arthur. Since the cruisers in Vladivostok were able to emerge (that port not being blockaded), the operation was not complete, and a danger of interference always existed.

This method of blockade is far more difficult to carry out in the present day, than formerly; owing to the existence of submarines and torpedo craft, the blockading ships have to remain further away from the port; there have to be inner lines of cruisers, scouts and destroyers; and quick concentration takes longer owing to the greater space covered by the blockading force, and more ships of all natures are required for the same reason.

Greater and more vigilance are required than in former days, because the enemy's ships can come out regardless of weather (thick weather would be their opportunity), and it is most important that not a single craft, from a battleship to a torpedo boat, be allowed to escape.

This method of blockade includes the commercial blockade, and all countries would be informed of its having been established.

(2) *Commercial Blockade.*—To prevent floating commerce from entering or leaving the blockaded harbour. The blocking force would not be powerful enough to prevent a squadron of battleships or cruisers from entering or leaving the port blockaded; and it would not be instituted outside a fortified military port, or one containing a strong naval force. But it would be able to stop scouts and torpedo craft from entering or emerging, unless in very great numbers; and if unable to stop them from emerging, would give warning of their escape and the direction in which they are going.

In both these forms of blockade it is usual, as a matter of courtesy, to allow neutral armed ships belonging to foreign navies to enter and leave for their own purposes, presumably connected with the subjects of their own country who are in the blockaded port. This, however, is not a right, and the country to which the blockading ships belong has a right to refuse it, and to back her refusal by force.

All countries must be notified of a properly instituted commercial blockade, in accordance with International Agreement.

(3) *Observing a Port.*—This, with its subsidiary operations, should be conducted in such a way as to induce the enemy to put to sea, the object of observing the port being primarily a naval one, viz., to bring him to decisive action.

The principal observing force (consisting of battleships and cruisers) would be either in one squadron, or more, provided that they were in supporting reach of each other, and so placed as to be able to cut off the enemy's fleet on emerging from the port observed before it can get dangerously near its probable objective, and yet sufficiently far out to ensure a battle before it can regain the shelter of its own ports. It is also worth noting that the battle should, if possible, be fought so as to make it difficult for the enemy's damaged ships to obtain the shelter of a friendly neutral's harbours before being captured.

The observed port must be watched closely, so that immediate notice of the enemy's exit may be given; and this would be done by small cruisers, scouts and destroyers, which should be strong and numerous enough to attack any torpedo craft trying to get to sea.

In order to induce the enemy's main force to put to sea it is important that every means be used to prevent his knowing that our fleet is observing the port, or if that be impossible, to do nothing which will lead him to suppose that his port is being observed.

This operation is not a blockade.

Subsidiary operations to induce the enemy's fleet to put to sea, may take the form of a diversion on the enemy's coast, or against some important part of his sea-borne trade, either by the observing fleet or by a force affiliated to it, or by any oversea movements calculated to interfere seriously with the enemy's war plan.

Concentration

The guiding feature of modern preparation for war is to be ready for rapid action. It is true at sea, more even than on land, that upon the first movements depend the initiative, the power of controlling the enemy's strategy, and of making him conform to our movements. This readiness for rapid action will depend on a proper distribution of the fleet so as to meet all the requirements.

The distribution of the fleet should be dominated by the idea of concentration, but it must be understood clearly what concentration means. Clausewitz says:—"The best strategy is always to be sufficiently strong, at first generally, then at the decisive point. There is therefore no higher or simpler law for strategy than this—keep your forces together."

The maxim "Keep your forces together" does not, however, necessarily mean keeping them all concentrated in one mass, but rather keeping them so disposed that they can unite readily at will. At sea it is more difficult than on land to foretell where the decisive point will be; but since it is quicker and easier at sea to concentrate forces at any particular point than on land, in applying this maxim for our purposes, the rule should be to dispose the forces at sea so as to be able to concentrate them in time at the decisive point so soon as this point is determined, and also so as to conceal from the enemy what it is intended to make the decisive point.

If the forces are rightly disposed within due limits, adequate control of all the lines of passage and communication can be assured, and if the enemy undertakes any operations it should be possible to ensure that sufficient forces can be concentrated in time to defeat his object. On the other hand, if the forces are concentrated in one mass, there can be little chance of deceiving or confusing the enemy, while it gives him an opportunity of successfully carrying out some operation by evasion.

The Peculiarity of Maritime Communications

Since the whole idea of command of the sea rests on the control of communications, it cannot be fully apprehended without a thorough understanding of the nature of maritime communications.

Ashore, the respective lines of communications of each belligerent tend as a rule to run more or less approximately in opposite directions, until they meet in the theatre of operations or the objective point.

At sea, the reverse is frequently the case; for in maritime warfare the great lines of communications of either belligerent often tend to run approximately parallel if, indeed, they are not identical.

Thus, in the case of a war with Germany, the object of which lay in the Eastern Mediterranean, or in America, or South Africa, our respective lines of communication would be identical.

This was also the case in all our imperial wars with France.

This peculiarity is the controlling influence of maritime warfare. Nearly all our current maxims of Naval Strategy can be traced to the pressure it exerts on naval thought.

It is at the root of the fundamental difference between Military and Naval Strategy, and affords the explanation of much strategical error and confusion which have arisen from applying the principles of land warfare to the sea without allowing for the antagonistic conditions of the communications and the operations against them in each case.

On land, the chief reason for not always striking the enemy's communications at once is that, as a rule, we cannot do so without exposing our own. At sea, on the contrary, when the great lines are common to both, we cannot defend our own without striking at the enemy's.

Therefore, at sea, the obvious opening is to get our fleet into such a position that it controls the common lines, unless defeated or evaded. This was usually done in our old wars with France, by our attempting to get a fleet off Brest before the French could sail.

Hence the maxims "That the proper place for our fleets is off the enemy's coast," "The enemy's coast is our true frontier," and the like.

But these maxims are not universally true; witness Togo's strategy against Rojesvensky, when he remained correctly upon his own coast.

Take, again, the maxim that the primary object of the fleet is to seek out the enemy's fleet and destroy it. Here, again, Togo's practice was the reverse of the maxim.

The true maxim is "The primary object of the fleet is to secure communications, and if the enemy's fleet is in a position to render them unsafe it must be put out of action."

The enemy's fleet usually is in this position, but not always.

Example.—Opening of the War of the Spanish Succession. The operations of 1702 were to secure some point (Cadiz, Gibraltar, or Ferrol) on the Spanish trade communications, the French lateral communications, and our own lines of passage to the Mediterranean, where was to be our chief theatre of operation. These last two lines were identical. In 1703, the chief operations had for their object to secure the alliance of Savoy, and particularly of Portugal. Rooke's official instructions directed that the French fleet was to be ignored unless it threatened the common communications. Result.—By 1704 we had gained a naval position from which France could not eject us, and she abandoned the struggle for sea communications.

But nine times out of ten the maxim of seeking out the enemy's fleet, &c., is sound and applicable—

(*a*) Because for us *general permanent command* is usually essential to ultimate success, and this cannot be obtained without destroying the enemy's fleet.

(*b*) Because usually the enemy's fleet opens with an attempt to *dispute the control of the common communications*.

(*c*) Because usually the functions of the fleet are so complex (*i.e.*, the calls upon it so numerous) that it will seek to strike a blow which will solve all the difficulties; *e.g.*, Sir. Palmes Fairborne's solution of the problem in 1703.

Also it must be remembered that nine times out of ten the most effective way of "seeking out the enemy's fleet" (*i.e.*, forcing an action on him) is to seize a position which controls communications vital to his plan of campaign.

This was what happened in 1704. Rooke was unable to seek out the Toulon Fleet, but by seizing Gibraltar he made it come to him (not intentionally, but by the operation of strategical law).

Practically all great naval actions have been brought about in this way, that is, they have been the outcome of an effort to clear essential communications from the enemy's fleet; *e.g.*, Gravelines, La Hogue, Quiberon, Trafalgar, Tsushima.

Similarly the great actions of the old Dutch wars were brought about because our geographical position placed us astride the Dutch trade communications, and they were forced to seek a decision against our fleet.

In applying the maxim of "seeking out the enemy's fleet" it should be borne in mind that if you seek it out with a superior force you will probably find it in a place where you cannot destroy it, except at very heavy cost. It is far better to make it come to you, and this has often been done by merely sitting on the common communications.

Again, if you seek out the enemy's fleet without being certain of getting contact, you may merely assist it in evading you, and enable it to get into a position on your own communications, from which it may be very costly to dislodge it. It was for this reason that the Elizabethan Government kept the fleet in home waters in 1588. Sampson, in the Spanish-American War, was actually permitted to make this mistake. By going to seek out Cervera without being sure of contact, he left him a clear run into Cienfuegos or even Havana, which it was the main function of the fleet to prevent. Captain Mahan has since modified this maxim as follows:—"Seek out the enemy's fleet, if you are sure of getting contact." A truer maxim would seem to be "Seek contact with the enemy's fleet in the most certain and favourable manner that is open to you." To seek out the enemy's fleet is only one way of doing this, and not always the best way. It must be remembered that other conditions being equal, it is an obvious advantage to fight in your own waters rather than in those of the enemy, and more likely to ensure that a defeat of the enemy will be decisive.

RN War College Portsmouth
January 1909

15 "REQUIEM FOR STRATEGIC PLANNING?"

(Selection from Chapter 12 of *The Accidental Admiral: A Sailor Takes Command at NATO*)

ADM James Stavridis, USN (Ret.)

In this thought-provoking piece, Admiral Stavridis, currently the dean of the Fletcher School of Law and Diplomacy at Tufts University, provides a real-world assessment of the strategic planning process, proposing that the vicissitudes of international relations, coupled with technological changes, have made traditional methods of strategic assessment and action too vulnerable and cumbersome. In contrast to what is normally anticipated from military thinking, he subordinates precision and accuracy to more practical expectations, allowing for proactive approximations rather than more methodical and comprehensive (and consequently slower) approaches. Cleverly utilizing some of the wisdom of ancient mythology to make his points, he makes clear the challenges that modern planners must face and he offers pragmatic advice, ultimately urging today's strategic planners to "write a plan, recognize that it needs to be produced quickly, and don't obsess over what you can't quite reach in terms of perfect information and assessment."

"REQUIEM FOR STRATEGIC PLANNING?"

(Selection from Chapter 12 of *The Accidental Admiral: A Sailor Takes Command at NATO*) by ADM James Stavridis, USN (Ret.), (Naval Institute Press, 2014): 149–55.

Everyone who has run a large organization (and most people who have run small ones) knows that planning is a key element of success. I have valued the process of strategic planning throughout my career, yet during my time at NATO I gradually began to lose my faith in the long-range strategic process. This was not due to any failure in our ability to produce coherent planning, but rather to the sense that we are departing the era of effective strategic planning and entering a very tactical world. Strategy is the big picture. It is the "what"—the overall goal that you are trying to achieve. Tactics are the "how"—the individual steps you take to get there.

Let me point out that I worked on strategic plans at every level in my nearly four decades of naval service, beginning with my first job as a division officer on a destroyer, where I crafted a long-range plan for the maintenance and inspection of our onboard nuclear weapons that were part of my responsibilities as the ship's antisubmarine warfare officer. Along the way, I wrote strategic plans for my destroyer, USS *Barry*; for the destroyer squadron I commanded in San Diego; for the big carrier strike group centered on USS *Enterprise*, a nuclear-powered aircraft carrier; and for both of my four-star U.S. commands—the Southern Command in Miami and the European Command in Stuttgart, Germany.

I have also been very involved as an analyst and drafter for numerous strategic plans ashore. Strategic planning was among my defined Navy sub-specialties, meaning that while I was primarily a warfighter at sea on destroyers, cruisers, and carriers, I was expected to have demonstrable skills in building a strategic plan. I learned many of these skills at the various War College courses I attended over the years, especially at the Naval War College and the National War College.

I mention all of this simply to make the point that such training is very much a part of any officer's kit bag, and I am no exception. Given my background in international relations, I particularly enjoyed the sort of very high level strategic planning that took into account the broader world. I had faith in our ability to craft long-range, far-seeing documents that laid out a coherent way forward. Over the course of my four years at NATO, however, the small slivers of doubt that had been creeping into my mind about this approach since 9/11 began to take root and sprout.

In beginning any strategic planning process, the first thing to consider is the environment itself. During my time at NATO the progression of events accelerated and the flow of information available to decision makers, which was high when I arrived, seemed to expand each year, growing as information technology grew. The ideas and concepts that we felt secure in planning toward felt outdated and inappropriate within a year or two, much like a laptop that falls further and further behind into obsolescence.

The traditional strategic planning model for a large military organization such as NATO is quite straightforward. It begins with such bedrock documents as the NATO Treaty, which is a model of elegance and brevity: twenty-four sentences in fourteen sections. The bedrock is supplemented by national guidance provided by key political decision makers—most recently, in the case of NATO, the strategic concept agreed to by the twenty-eight heads of state and government who met at the Lisbon summit in 2010. The Lisbon agreement defines key missions for the alliance and includes direction on building partnerships, establishing cybercapability, seeking a "true strategic partnership with the Russian Federation," and other tasks.

Armed with bedrock documents and national guidance, a planner ideally would conduct an assessment of the overall security environment and the progress on previous plans—taking a kind of navigational fix on the present. This should include some outside sources—both classified and unclassified—as well as the personal views of the commander. I liked to let my planners come and visit with me every couple of weeks during the strategic planning cycle so that I could continuously inject my ideas and views into this critical first assessment phase—about both the environment and where we stood at the moment.

This part of the cycle is where the first real difficulties in traditional strategic planning occurred. It is hard to assess an environment when things are changing rapidly, and certainly that was the case during my time at NATO. The obvious huge muscle movement was the Arab Spring (with attendant explosions in Libya,

Egypt, and Syria). These revolutions quite literally reshaped the politics of the southern and Levantine borders of the alliance. We also saw the global economy lurch into the so-called Great Recession; the collapse of several economies on the southern European tier of the alliance; the failure of relations with Russia following the election of Vladimir Putin; Iran's accelerated pursuit of nuclear weapons; a constant shift in the tone and approach in Afghanistan; the disclosures of Edward Snowden, which more or less unglued the cyberworld; the dramatic reshaping of the global energy calculus as a result of shale gas exploitations; the expansion of social networks (Facebook and Twitter) to become the third and fifth largest "nations" in the world; and a thousand other radical changes. The old programmers' maxim "garbage in, garbage out" occurred to me as I contemplated our strategic planning process. If we didn't have a clear assessment of the environment and our place in it, and if things were changing every five minutes, what could we do?

The traditional planning process moves from the assessment to guidance development by the key leadership team and the production of a theater campaign plan—in layman's terms, a list of things you hope to accomplish. Sometimes strategic planning is boiled down to the real essentials: ends, ways, and means. The development of a campaign plan is the "ends," or simply where you want to end up. In the case of NATO, we leaned heavily on the strategic guidance, which meant that we needed to be ready to defend the alliance (through exercises and training), reach outside the alliance to build partnerships (as in Afghanistan, where the twenty-eight NATO nations are joined by twenty-two partners), conduct effective crisis management (a Libya-like operation), and create at least amicability with Russia (getting harder and harder).

Once you have an assessment and a campaign plan, it is a relatively simple matter to decide on the capabilities you need to accomplish your goals—the "ways." Simply put, how are you going to do this? In the case of NATO, we could see that we needed more and better intelligence and surveillance (especially from unmanned platforms); more special forces; vastly improved cybercapability; stand-off strike weapons; better-honed strategic communication and legal skills; and a variety of other things.

During my time at NATO we were able to make some progress on most of that list, despite declining defense budgets. We built up the NATO special forces headquarters and brought online the new Global Hawk airborne unmanned vehicle, which has a terrific wide-area surveillance capability. As a result of the Libyan operation, we recognized the need for a variety of other "ways" to do business and tried to adjust the alliance's approach accordingly.

In all of this we were working closely with the supreme allied commander for transformation, General Stéphane Abrial of France. A tall, rangy, good-natured fighter pilot by trade, Stéphane was a superb partner in the world of strategic planning. He had an organization of about a thousand officers and enlisted personnel from the twenty-eight member nations who worked very hard to create more efficient ways to accomplish the mission under constantly changing operational, political, economic, and diplomatic conditions. His command, based in Norfolk, Virginia, had the lead on "ways" and did a good job—especially with the NATO Smart Defense initiative, which sought to rationalize the ways (and ultimately the means) of our strategy.

Which brings us to the third simple leg of strategic planning: "means," or the resources to get the job done. Most obviously, these resources are the defense budgets of the nations. Here there is both good and bad news, as is usual in strategic planning. It is a given that there is never enough funding to buy a military everything it wants—think of that as the iron rule of military strategic planning. But the good news is that NATO is an incredibly rich alliance. It represents more than 50 percent of the world's gross domestic product and spends more than $900 billion annually, about two-thirds of that from the United States and the rest from the other nations. Ideally that ratio of U.S. to non-U.S. spending would be roughly fifty-fifty, which would correspond to the GDP shares involved. But given that no non-NATO nation is spending more than $150 billion (China is at the top) annually on its military, it is clear that NATO has plenty of resources to apply to this part of strategic planning. I was frustrated by the declining European defense spending (now mirrored in the United States as well), but it seems to me that we nevertheless have enough to defend the alliance adequately.

To summarize, the four elements of the classic strategic planning process are assessment and environment, guidance, capability determination, and resources—or, to simplify, ends, ways, and means. I have used them in many situations to develop solid strategic plans, ranging from the enormous Quadrennial Defense Review that guides the Pentagon every four years through a $600 billion budget and global operations involving millions of military and civilians to the tiny plan I made for the hundred sailors working for me as an operations officer on a cruiser in the mid-1980s.

But twenty-first-century planning seems to be departing the age of rational strategic planning and entering a relentlessly tactical period. Beyond the acceleration of knowledge and events that I discussed earlier, there are several other factors at work here, which can be illustrated with three characters from Greek mythology.

The first is Tantalus, whose name gives us the English word "tantalize." The Greek gods punished Tantalus for misbehavior by chaining him to an apple tree in Hades in water up to his waist and inflicting him with overpowering thirst and hunger. Unfortunately for Tantalus, every time he bent over to assuage his thirst, the water receded from his lips. Likewise, when he reached up to pluck an apple to satisfy his hunger, the wind would blow the branches just out of his reach. The same problem tends to manifest itself in strategic planning today: planners work too hard to write the perfect plan, only to see the effort drift just beyond their reach. With just a little bit more effort, a better and more perfect word, a little more time, we are certain we can finally write the perfect plan. Forget it. To paraphrase George Patton, a good (read "more tactical") plan executed now is better than a perfect plan that arrives next week (or never arrives at all, like the apples and water for Tantalus). The lesson here is write a plan, recognize that it needs to be produced quickly, and don't obsess over what you can't quite reach in terms of perfect information and assessment.

The second Greek character, Sisyphus, is perhaps a bit better known. He is the fellow who is forever rolling a boulder up a hill in Hades. Just as he gets it to the top, it crashes back down and he has to start over again—forever. It's a very apt metaphor for the outcome of many of our plans. We craft them very carefully, assess them in minute detail, figure out how to accomplish our goals, put resources behind them, get ready to execute—the boulder is just at the top of the hill—and suddenly there is a great lurch and the plan has to be changed. Three good recent examples include 9/11 and the rise of global jihad, the Arab Spring across much of the Levant, and the emergence of shale gas as an energy source. All three are having and will continue to have dramatic effects on the global scene.

The third Greek I would introduce is Prometheus, who provided humankind with fire against the wishes of the gods. Zeus had him chained to a rock and each day sent an eagle to eat his liver. Those who provide full transparency in strategic planning suffer a similar (although far less bloody) fate. And in today's world, *everything* has essentially become transparent. Most governments cannot protect even the most sacrosanct secrets and must assume that any strategic plan will be exposed to the light of day, to be picked apart by critics, adjusted by adversaries, and subverted by entrenched entities whose interests it gores. In other words, the plan's drafters (and those who try, generally fruitlessly, to execute it) are treated to a daily liver extraction.

The moral of these not-so-amusing Greek myths is not that strategic planning is an endless form of torment, although to some it may seem that way. The point is that the pursuit of perfection, the potential for sudden catastrophic change, and the ill effects of forced transparency that the myths illustrate have made strategic planning in this brave new world grueling, frustrating, unending, and of less use than it once was. Couple that with the accelerating rush of information and constantly changing tactical events, and it appears to me that we are witnessing at least a temporary requiem for grand levels of strategic planning, dear though they are to the military's heart.

What should we do instead?

A better approach, it seems to me, would drop the more grandiose planning efforts and instead start with a very broad guess at the five-to-ten-year future, recognizing that the prediction won't be very accurate. It would then move swiftly to a simple set of long-term goals in very general terms about the organization's desired direction, speed, and distance to travel—in the Navy, we would call these sailing directions. The real focus of planning in today's world should be on a detailed annual planning process that responds to the goals and the general sailing directions with tactical judgments.

At NATO, we would look very broadly each year at where the organization was supposed to be headed in general terms and then craft specific annual goals to keep us moving in that direction. An example would be Afghanistan: the broad goal was to turn over security operations to the Afghans and withdraw the NATO and coalition forces over the next five to ten years. Each year we would work very hard defining the specifics of what we hoped to accomplish that year: how many Afghan police trained, how many Afghan infantry soldiers, how many Afghan special forces, how much logistics capability (trucks, planes, fuel) purchased, how many coalition bases handed over to the Afghans, how many children in schools, and so on. It was a very tactical set of annual goals, and we constantly measured our progress as the year went on. At the end of the year we would craft another set of annual goals.

Innovation is key in such tactical planning. It is easy to slip into repeating goals from year to year without injecting sufficient thought into change and new ideas. Having an innovation cell is critical, and that group should be a big part of the goal-setting process because they can think outside the box and help provide new directions to be undertaken in a given year.

Naturally, there will be some projects that have long-term components. But such projects should be taken on with a healthy respect for the fallibility of our predictions and recognition that today's world is far more tactical given the scope and pace of change. We should still have long-range goals in mind, but they should be general in nature and flexible to adjustments as we sail on.

The Greeks have an expression that translates loosely as "man plans, fate laughs." Expect plans to fail, even the tactical ones. Eisenhower once said, "In preparing for battle, I have always found that plans are useless, but planning is indispensable." I believe he meant that the teamwork, bonding, hard intellectual work, and goal setting involved in preparing a plan—be it strategic or tactical—pay off when events come fast and furiously. Never has that been truer than today. We are constantly bombarded with far more information than we can analyze, and detecting nuances is increasingly difficult.

My advice, based on four years in NATO, is this: plan with an eye toward the tactical and the annual horizon; recognize that the sailing directions for the long-distance voyage are going to be frequently revised as the wind and seas change; and, above all, keep the ship sailing forward with purpose and do not allow yourself to merely drift before the elements on an uncaring sea.

INDEX

Abrial, Stéphane, 187
access denial and control, 143–45, 146–49, 149n2
Adams, David A., 151–60
Afghanistan, 186, 189–90
Africa, 71–72
air attack (presence force action), 20
Air Force, U.S.: Air Expeditionary Forces, 146; airfield protection responsibilities of, 41; functions and operations like a navy, 141, 145; Global Reach, Global Power doctrine, 46; globalization and access control, 148–49; national security policy and role of, 34; opinions about Navy, 27–28; public perception and opinions about, 28
air power: carrier-based air power and projection of naval power, 39–40; carrier-based air power and sea control, 12, 19; mechanization and, 132; naval air power role in Korean War, 4; projection of power ashore and naval tactical air, 15–19, 23; tactical air tactics, 16–18; theories of strategy for, 136–37
air strategy, 122, 135–36, 153
aircraft carriers: carrier-based air lifts, 39; carrier-based air power and projection of naval power, 39–40; confusion over use of, 19; power projection, presence, and, 7–9, 19; sea basing of Navy and, 40–41; sea control and carrier-based air power, 12, 19; strategy for carrier groups, 114–15
Air-Land Battle doctrine, 46, 50
Air-Sea Battle, 112, 118
alliances and coalitions: national security policy and development of, 34. *See also* global maritime coalition and Free World navies; North Atlantic Treaty Organization (NATO)
amphibious assault operations: presence force action, 20; projection of power ashore and, 15, 23, 39–40; role of Navy in, 30
anti-aircraft fire (AAA), 16, 17, 18
antisubmarine operations, 41, 68, 70
antisurface warfare (ASUW) operations, 68–69
Arab Spring, 185–86, 188
Arabian Gulf, 72
Argentina, 66, 114–15
armies, function of, 143–45, 146–47
arms and revolution, 130, 134
Army, U.S.: Air-Land Battle doctrine, 46, 50; guard responsibilities of, 41; national security policy and role of, 34; public perception and opinions about, 100–101; World War I and personnel for, 98–101
assured second strike capability, 21, 22

Atlantic Ocean: asserting right to use some seas and reinforcement of Europe, 13; contributions of global maritime coalition to security in, 69, 71; sea control and naval strategy during World War I, 12, 85
Australia, 70, 72

Bacon, Francis, 77, 144
Barry, 184
Base Force concept, 46, 51
Belgium, 69, 90–91
Black September, 8
blockade: close blockade, 176–77; commercial blockade, 177; presence force action, 20; role of Navy in, 30; sortie control and sea control, 13; World War I operations, 12
bombing operations: presence force action, 20; projection of power ashore and naval bombardment, 15; World War II bombing of Japan, 151–52, 154–55, 160n12
Brazil, 71
Breemer, Jan, 150n5
budgets and funding: expenses of Navy, public perception about, 3; funding for Navy, 33; NATO alliance resources and budgets, 187; naval forces and cost of oceangoing navy, 142, 145–46; naval strategy, understanding of and budget decisions, 4; reduction of, popular desire for, 28; strategic planning and, 187–88
Byron, John L., 45–52

Cape Matapan, Battle of, 12, 19
Castex, Raoul, 102, 106, 109
Center for Strategic and International Studies, "Maritime Security Dialogue" conference, 2, 65
checklists, 1–2
Cheney, Dick, 49
China: *CS–21* and, 58, 60; naval forces of, 142; naval influence of navy, 105–6; sea-denial capabilities of, 113; strategy for operations in littorals adjacent to, 112; trading partners and threat of war with, 76–77, 79, 81–82

choke points: amphibious assaults on, 18; contributions of global maritime coalition to security of, 68, 69; control of and sea control, 13, 14; Soviet control of and commerce, 68
Civil War, 30, 31, 98, 99
Clausewitz, Carl von, 24, 99, 153–54, 157, 159n10, 166, 167, 178
close air support, 17
close blockade, 176–77
Cold War: continental threat and maritime strategy, 84; decisive battles of, site of, 38; naval doctrine and the transoceanic navy, 34–44; winning of, 158
Colin L. Powell Joint Warfighting Essay Contest, 141, 151
Colombia, 71
colonialism, sea power and, 35, 36
combat-credible forces, 115–17
command of the sea. *See* sea control/command of the sea
commerce and trade: global maritime coalition and security of, 66, 70, 71, 73; sea control, security of movement by sea, and, 11; sea lanes of trade, Navy role in keeping open, 4, 30; Soviet control of choke points and, 68; trading partners and threat of war with, 76–77, 79–80, 81–82, 86; U.S. as largest trading nation, 86
commercial blockade, 177
communications, maritime, 171–72, 179–82
concentration of fleet, theory of, 32–33, 36, 42–43, 178–79
containment, 61–62, 108, 144, 156, 158
Continental Phase of national security policy, 29–31
continental strategy, 122, 135–36, 153
A Cooperative Strategy for 21st Century Seapower (*CS–21*): adaptation of to conditions and circumstances, 61–63; audience for, 61; coherent tie between strategy and goals, 59, 60; core capabilities in, 7; core challenges and, 56, 58–59; fleet design and, 55, 57–58, 59; goals and objectives, 59; implementation of, 61; negative

comments on, 55; opportunity-based compared to competitive oriented strategy, 59; prevention of war, importance of, 115, 117; purpose of, 57–58; updating of, 54, 62–63
Corbett, Julian S., 102, 106, 109, 161–82
counter-air/anti-air operations, 16, 17–18
Cuban Missile Crisis, 4
Cutler, Thomas J., 7–9
cybercapabilities, 186
cyberworld, 186

Dardanelles campaign, 77, 78–79, 92
deception, 14
decisive action, site of, 37–38
decisive force: advocates for, 153, 160n27; appeal of concept, 157; joint doctrine and, 153, 157–59; myth of and problems with, 157, 158–59; naval power compared to, 151, 157–59; theory of, 152–53; victory through use of, 152–53
defensive strategy: advantages of, 165; disadvantages of, 165; functions and characteristics of, 165–67; negative object and, 164–65; offensive operations with defensive intentions, 167–68; relation of offensive to, 165
Dempsey, Martin, 62
deployment tactics: preventive deployment, 19; reactive deployment, 19
deterrence: assured second strike capability, 21, 22; balance of power and strategic attack force capabilities, 21–22; concept of and evolution of mission, 20–23; deterrent missions of Navy, 19, 117; naval quarantine and role of, 4; objectives of strategic deterrence, 21; power projection and, 22–23; presence as deterrent to war, 20–21; sea control and, 22–23; strategic deterrence, importance of understanding objectives of, 10
diplomacy, gunboat, 19

East Asia, combat-credible forces for, 115–16
economics and the economy: contributions of global maritime coalition to security and, 73; globalization and, 147–48; sea control, security of movement by sea, and, 11; trading partners and threat of war with, 76–77, 79–80, 81–82, 86
Egypt, 69, 186
Eisenhower, Dwight D., 190
enemy's fleet, seeking out, 180–82
energy, shale gas exploration and global shift in, 186, 188
Enterprise, 184
Eurasian Phase of national security policy, 34
Europe: defense spending in, 187; distribution of sea power in, 35–37; domination by single power, threat of, 138–39; geographical-naval contributions of global maritime coalition in, 68–69; power projection from Mediterranean and operations in, 42
explosives in war, 130, 132–34

Falkland Islands, 114–15
fleet: combat-credible forces for, 115–17; deployment of and theory of concentration, 32–33, 36, 42–43, 178–79; dispersion and mobility of, 42–43; enemy's fleet, seeking out, 180–82; force strength of, 27; maritime strategy and design of, 55, 57–58, 59; mobility of, 115, 117; strength of battlefleet and shipbuilding races, 32; unlocated/untargeted, maneuvering to stay, 114–15
flight and warfare, 130, 136–37
flotilla operations: being there and use of, 117–19; feasibility of, 119; strategy for using, 111–19
forward presence. *See* presence/forward presence
France: global maritime coalition contributions of, 69, 70, 72; Napoleon and Napoleonic War, 35, 82, 84, 98, 127–29, 134, 157, 175; naval competition and sea power of, 35; naval forces of, 142; occupation of Vietnam by, 155–56; World War I and coalition with Britain, 78–79, 94

Free World navies. *See* global maritime coalition and Free World navies
Friedman, Norman, 4, 76–86
Frothingham, Thomas G., 87–101

Garrett, H. Lawrence, 52
general sea control, 106, 174, 176, 181
geographical focus of operations, 41–42
Germany: ASUW operations, role in, 68–69; fleet for World War I, 83, 89–90, 92; internal crisis and World War I, 80–81; naval competition and sea power of, 35; navy of, 27; sea control and naval strategy during World War I, 11–12; threat from navy, U.S. concern about, 33; World War I and relationship with Britain, 76–77; World War II planning for use of submarines, 12
Global Hawk unmanned vehicle, 187
global jihad, 188
global maritime coalition and Free World navies: conferences to discuss ocean issues, 74; contribution of security and defense, 65–66, 75; disputes between coalition members, 66; exercises and joint operations, 66, 73, 74, 75; geographical-naval contributions, 67–72, 74, 75; linkage development, 74; political-economic contributions, 72–73, 74, 75; purpose of, 65–66; recommendations to enhance positive interaction in coalition, 73–75; structure of the coalition, 65, 66; technology transfer between, 75
globalization: access as key to, 147–48; concept of, 147; jointness concept and, 142–43, 148–49
"the gouge," 1
Gray, Colin, 62
Great Britain: army of, strength of, 78; Falklands dispute, 66, 114–15; global maritime coalition contributions of, 68, 69, 72; Napoleon and Napoleonic War, 35, 82, 84, 127–29, 157; naval competition and sea power of, 35; naval forces of, 142; sea control and carrier-based air power, 12; sea control and naval strategy during World War I, 11–12; sea power and sea control of, 35, 77, 127–29, 140; World War I, entrance into and character of war, 77, 82–84; World War I and coalition with France, 78–79, 94; World War I and relationship with Germany, 76–77; World War II planning for use of submarines, 12
Greece: communist threat to and assistance from U.S., 138–39; global maritime coalition contributions of, 69; Greek myths as explanation of strategic planning limitations, 188–89
Greenert, Jonathan, 2–3, 54, 65
Gulf War, 152
gunboat diplomacy, 19

Hattendorf, John, 120
history: looking back at mistakes of the past, 76, 77, 86; no repeats of, 76, 77
Hoffman, Frank, 53–64
Huntington, Samuel P., 3, 24–44

Independence, 7–8, 114
Indian Ocean, 70, 72
information warfare and access to information, 147–48
inner jointness, 48–49, 50–51
innovation, tactical planning, and innovation cells, 190
intelligence-gathering and sharing operations, 72
interdiction tactics: battlefield interdiction tactic, 16–17; deep interdiction tactic, 16
International Maritime Organization, 66, 73
intimidation, sea control and, 15
Iran: global maritime coalition and response to threat from, 66; nuclear ambitions of, 186; sea-denial capabilities of, 113; strategy for operations in littorals adjacent to, 112
Iraq: global maritime coalition and response to threat from, 66; power projection, presence, and Jordanian Crisis, 8
Israel, 8, 69, 72, 115

Japan: basing of carrier group in, 146; global maritime coalition contributions of, 70, 72; naval competition and sea power of, 35; navy of, 27, 37, 142; surrender of, 152, 155; threat from navy, U.S. concern about, 33; U.S. presence in, 155; World War II bombing of, 151–52, 154–55, 160n12; World War II plan for use of submarines, 12; World War II war plan, 154–55
jointness: assignment of roles by function, 141; assignment of roles by service, 141, 143; benefits of, 87; decisive force doctrine and, 153; functional jointness, 141; globalization and concept of, 142–43, 148–49; inner jointness, 48–49, 50–51; naval approach to, 153–54; sea control, unity of command, and, 107
Jordanian Crisis, 3–4, 7–9
Junge, Michael, 54
Jutland, Battle of, 11–12, 94

Kaplan, Robert, 57–58
Korean War: land battles during, 38; naval air power role in, 4; naval bombardment and projection of power ashore, 15; naval tactical air operations during, 16; public understanding of operations, 4; sea control and, 4
Kosovo, 152

Law of the Sea Conference and treaty, 66, 73
Leyte Gulf, Battle of, 3
Libya, 66, 185–86, 187
limited war, 169
littoral combat ships (LCS), 112, 116–17
littoral/coastal areas: coastal defense responsibilities, 30, 32; combat-credible forces for, 115–17; dispersion of fleet for patrol operations in, 112; flotilla operations for, 111–19; maritime strategy for, 112–19; mechanization and operations in, 131–32; naval power in, application of and weapons for, 38–40, 41; sea-denial capabilities of China and Iran in, 113; ships for use in, 112; as site of decisive actions, 37–38
local defense, 14
local sea control, 106, 174, 176
logistic support, 130, 134–36
Luce, Stephen A., 36, 120

Mahan, Alfred Thayer: command of the sea doctrine, 32–33; maritime power, definition of, 11; police function of Navy, opinion about, 32; relevance of, 24, 102, 109; sea power and naval warfare strategy of, 35, 36–37, 38–39, 157–59; strategic concept for the Navy, 32–33
Major Strategy, 162–64
Marine Corps, U.S., 141, 143
maritime communications, 171–72, 179–82
"Maritime Security Dialogue" conference, Center for Strategic and International Studies, 2, 65
Maritime Strategy: as alternative way to fight continental war, 78; competitive orientation of, 60; development of, 47; enduring theme of, 49–50; ownership of, 50–51; simplicity of, 47; success of, 46; theme of, 49; timing of, 51
Marne, Battle of the, 91
mechanization in war, 130, 131–32
Mediterranean Sea: national security policy and naval forces in, 42; naval complexes in and security in, 72; naval confrontation in, 114; power projection from, 41–42; power projection, presence, and aircraft carriers, 7–9
Mexican War, 29, 30
Middle East: combat-credible forces for, 115–16; geographical-naval contributions of global maritime coalition in, 69–70; power projection from Mediterranean and operations in, 42; power projection, presence, and crisis in, 7–9
Midway, Battle of, 3

military services: elements of, 25–26; national policy and capabilities of, 25–26; naval operations, 143–45; opinions about Navy, 27–28; organizational structure of, 26; postwar reaction to, 28; public perception and opinions about, 25–26, 100–101; relationship with citizens, 88; strategic concept and role of, 25–26
minelaying and mine clearance operations, 69, 72
Minor Strategy, 162, 163
Mintzberg, Henry, 54–55
missiles: guided missile development and projection of naval power, 39–40; sea-based cruise missiles, 146–47; surface-to-air missiles (SAMs), 16, 17, 18
Monroe Doctrine, 29
Murray, Williamson, 54

Napoleon and Napoleonic War, 35, 82, 84, 98, 127–29, 134, 157, 175
National Military Strategy, 46, 52, 153
national security policy: capabilities of Navy and, 23, 25–28; Continental Phase and role of Navy, 29–31; Eurasian Phase and role of Navy, 34; international balance of power, threats to security, and, 26; naval forces in Mediterranean basin and, 42; nuclear weapons threats and, 34; Oceanic Phase and role of Navy, 31–34, 43; police function of Navy, 32
National War College, 184
naval doctrine and the transoceanic navy, 34–44
naval influence: concept of, 105–6; forward presence and, 105; peacetime and, 105
Naval Institute: Prize Essay Contest, 120; *Proceedings*, 2; strategy, articles and publications about, 2, 4–5
Naval Operations, Office of, 33–34
naval power. *See* sea power/naval power
Naval War College: strategic planning coursework at, 184; Strategy and Policy course at, 161; Strategy and War course at, 1–2; teaching methods at, 1–2

navies: access control by, 143–45; definition of, 143–45; demise of navies, 142, 145–46, 149; globally deployed navy, U.S. Navy as, 141, 142, 146; naval forces and cost of oceangoing navy, 142, 145–46; naval warfare and one navy, 142
Navy, U.S.: base of, 40–41; crisis of and questions about reason for existence of, 26–28; deterrent missions, 19; exercises and training with Free World navies, 73, 75; importance of, understanding of, 3; as last globally deployed navy, 141, 142, 146; missions of, 2–3, 10–11, 22–23, 38–40, 143 (*see also specific missions*); national security policy and capabilities of, 23, 25–28; national security policy and role of, 29–34; naval doctrine and the transoceanic navy, 34–44; organizational structure of, 33–34, 43; public perception and opinions about, 3, 24, 25–28, 33, 43–44, 100–101; reason for existence, communication of, 4, 11, 43–44; strategic concept and role of, 26–28, 32–33; success of and image of, 3; warfighting missions, 19; World War I and personnel for, 98–101
Navy Department, 33
Navy League, 33
negative object, 164
Netherlands, 35, 125
New Zealand, 70, 72
Nimitz, Chester W., 38, 39, 40–41
North Atlantic Treaty Organization (NATO): formation of, 139; geographical-naval contributions of global maritime coalition, 68–69; maritime coalition of, 66, 139; maritime strategy and, 139–40; mission of, 139–40, 185; naval exercises conducted by, 146; organizational structure of, 139–40; Smart Defense initiative, 187; strategic planning of, 185–87; wealth of alliance, 187
North Korea, 66, 70
nuclear weapons: bombing of Japan, 151–52, 154–55, 160n12; maritime

strategy and, 130, 137–38; national security policy and threat from, 34; partial nuclear attack, prevention of, 21–22; third powers, deterrence of, 22

object, nature of, 164–65
observed port, 177–78
Oceanic Phase of national security policy, 31–34, 43
offensive strategy: advantages of, 165; counter attacks, 167–68; disadvantages of, 165; diversions, 168; offensive operations with defensive intentions, 167–68; positive object and, 164–65; relation of defensive to, 165
Okinawa, 3
operations: geographical focus of, 41–42; open area operations, 14; science in, 1

Pacific Ocean: contributions of global maritime coalition to security in, 70, 72; focus of naval operations in, 41, 42; Mediterranean as replacement of focus of naval attention, 42; naval complexes in and security in, 72
Panetta, Leon, 62
patience, 151, 158
Patton, George, 188
permanent sea control, 107, 174–75, 176, 181
Persian Gulf, 61, 68, 69, 72, 117
Philippine Islands, 70
Philippine Sea, Battle of the, 37
piracy, 11
planning: approximations and broad goals over perfect information and assessment, 183, 188–90; failure of plans, expectation of, 190. *See also* strategic planning; tactical planning
Portugal, 69
positive object, 164
Powell, Colin, 49, 51–52, 160n27
power projection: amphibious assault operations and, 15, 23, 39–40; beyond the sea, 141, 148–49, 149n1; carrier-based air power and projection of naval power, 39–40; concept of and evolution of mission, 15–19, 23; deterrence and, 22–23; forward presence and, 9, 105; maritime strategy and, 122, 124–26; naval tactical air tactics, 15–19, 23; presence, aircraft carriers, and, 7–9, 19; projection of power ashore, importance of understanding objectives of, 10; sea control relationship to, 18–19, 104–5, 108; strategic success through, 157–59
presence force: actions available to, 19–20; composition and size of, 19–20; selection of and perception of countries, 20
presence/forward presence: concept of and evolution of mission, 19–20, 23; deterrent effect of, 20–21; importance and priority of, 2–4; naval influence and, 105; objectives of, importance of understanding, 10; power projection, aircraft carriers, and, 7–9, 19; power projection and, 9, 105; strategic success through, 151, 157–59; tactics for, 19; understanding of, 3–4, 20; U.S. presence in Japan, 155
primary objects, 162
Prometheus, 189
Putin, Vladimir, 186

Quadrennial Defense Review, 187–88

reconnaissance, exposure through, 20
revolution and arms, 130, 134
Rio Pact Alliance, 66, 70–71
Royal Naval War College, 161
Rubel, Robert C. "Barney," 111–19
Rumelt, Richard, 55, 56, 57, 58–61

sailing directions, 189, 190
science and scientific method, 1–2
sea control/command of the sea: achieving, tactical approaches for, 13–14; air, surface, and subsurface control, 106–7; asserting right to use some seas, 13, 112–13; carrier-based air power and, 12, 19; concentration of fleet, 32–33, 36, 42–43, 178–79; concept and definition of, 103, 108–9, 172–74; concept of and evolution of mission, 11–15, 23; as core capability, 104–5, 143–44; degrees of control, 106–7; denying

an enemy the right to use some seas, 13, 112–13; deterrence and, 22–23; dispute of command, 175–76; general, 106, 174, 176, 181; limitations on, 12–13; local, 106, 174, 176; Mahan command of the sea doctrine, 32–33; maritime communications, 171–72, 179–82; maritime strategy and, 122–26, 139–40; methods of securing control, 176–79; objectives of, importance of understanding, 10; obtaining, maintaining, and exercising, 107–9; offensive objective of, 103; passive techniques for, 14–15; peacetime and, 105, 174; permanent, 107, 174–75, 176, 181; power projection relationship to, 18–19, 104–5, 108; purpose of, 103–4; sea denial compared to, 102–3, 104; security and, 3–4; submarines and, 12; temporary, 107, 174–75, 176; understanding of, 3–4, 102–3, 109; unity of command and joint operations for, 107

sea denial: capabilities of China and Iran, 113; defensive objective of, 103; sea control compared to, 102–3, 104; understanding of, 102–3, 109

sea power/naval power: distribution of, 35–37; effectiveness of compared to decisive force, 151, 157–59; goals of, 147; in littorals, 38–40, 41; Mahan naval warfare strategy and, 35, 36–37, 38–39, 157–59; maritime strategy and, 122–26, 139–40; preservation of, 2; security and, 3–4; utilization of, 2

sea-based cruise missiles, 146–47

security: international balance of power and threats to, 26; naval power and, 3–4; sea control and, 3–4; strategic planning and transparency of plans, 189. *See also* national security policy

September 11 attacks, 188

Sherman, Forrest, 46, 47, 48

Sinnreich, Rick, 62

Sisyphus, 188

Sixth Fleet/Sixth Task Fleet, U.S., 7–9, 42, 114, 115, 140

Smart Defense initiative, 187

Snowden, Edward, 186

social networks, 186

Some Principles of Maritime Strategy (Corbett), 161

Somme, Battle of the, 85, 94

sortie control, 13

South America, 70–71

South Korea, 70, 72

Soviet Union/Russia: ASUW operations against, 68–69; balance of power and strategic attack force capabilities, 21–22; base expansion by, 72; commerce and control of choke points by, 68; domination by, threat of, 138–39; exercises with NATO naval forces, 146; global maritime coalition and response to threat from, 66; naval confrontation with U.S., 114; navy of, 27, 142; partial nuclear attack, prevention of, 21–22; power projection from Mediterranean and operations in, 42; power projection, presence, and Jordanian Crisis, 8–9; presence force and perception of, 20; presence force and perception of allies of, 20; relationship with U.S., 186

Spain, 35, 37, 125

Spanish-American War, 31–32

Stavridis, James, 65–75, 183–90

strategic planning: assessment and ends, ways, and means process, 183, 185–88; budgets and resources for plans, 187–88; campaign plan development, 186; faith in and doubts about, 184, 185; Greek myths as explanation of strategic planning limitations, 188–89; importance of, 184–85; pace of events and, 185; traditional assessment and action, 183; training in, 184; transparency and, 189

strategic thought: capacity for, 54; culture of, 54; decline in competence for, 62–63

strategy: adaptation of to conditions and circumstances, 54–55, 56, 61–63; audiences for, 59, 61; bad strategy, elements of, 55; complexity of and approach to teaching, 1–2; constraints and restraints on, 58; core challenges and, 56, 58–59; defensive strategy, 164–65; goals and objectives and, 55, 58, 59; good strategy, elements of, 55–56; Major Strategy, 162–64; meaning of concept, 53–54; Minor Strategy, 162, 163; offensive strategy, 164–65; pitfalls or sins, 58–62; process for formation and implementation of, 54. *See also* defensive strategy; offensive strategy

strategy, maritime/naval: approaches to study of, 121; benefits of, 57; budget decisions and understanding of, 4; contemporary use and patterns, 138–40; definition of, 172–74; development of, 2; end of naval strategy, 149, 150n5; examples of use of, 126; factors that complicate modern strategies, 130–38; fleet design and, 55, 57–58, 59; importance of, 1–5; logistic support and, 135–36; as Major Strategy, 163–64; purpose of, 57–58; revision of, 2; theory of, 121–26; understanding of, 2. *See also A Cooperative Strategy for 21st Century Seapower (CS–21)*; Maritime Strategy; vision statement/message

submarines: antisubmarine operations and global maritime coalition, 68, 70; antisubmarine warfare and mission of the Navy, 41; Polaris/Poseidon/Trident forces, 21; sea control and naval strategy during World War I, 12; World War I submarine warfare, 84–85, 92, 94–95, 96, 97; World War II planning for use of, 12

Suez Canal, 68, 69, 72
Sun Tzu, 8, 151, 158
surface-to-air missiles (SAMs), 16, 17, 18
surveillance operations, 68
Swartz, Peter M., 45–52
Syria, 7–9, 66, 186

tactical planning: Afghanistan goals and tactical planning, 189–90; failure of plans, expectation of, 190; innovation and, 190; shift in planning toward, 184, 188–90

tactics, science in, 1
Tangredi, Sam J., 141–50
Tantalus, 188
temporary sea control, 107, 174–75, 176
territory, control of, 143–45
Thatcher, Margaret, 115
theatre of operations, 169–70, 171
theatre of war, 169
Third Nations, presence force and perception of, 20
Tibbetts, Paul, Jr., 151
trade. *See* commerce and trade
transoceanic navy and naval doctrine: base of the Navy, 40–41; concept of the transoceanic navy, 34–35; decisive action, site of, 37–38; geographical focus of operations, 41–42; international power, distribution of, 35–37; mission of the Navy, 38–40; naval tactics, 42–43; public perception and opinions about the Navy and, 43–44

trip wire, 116–17
Tritten, Jim, 49
Turkey, 42, 69, 138–39
Turner, Stansfield, 10–23

ulterior objects, 164–65
United States (U.S.): balance of power and strategic attack force capabilities, 21–22; defense spending in, 187; moral force for World War I, 98–100; naval competition and sea power of, 35; naval confrontation with Soviets, 114; relationship with Russia, 186; World War I, entrance into, 84–85, 88–89, 95–101; World War II planning for use of submarines, 12. *See also* national security policy
unlimited war, 169

Vego, Milan, 102–10
Venezuela, 30, 71
Vietnam: French occupation of, 155–56; global maritime coalition and response to threat from, 66; noncommunist government in, preservation of, 156
Vietnam War: defeat of U.S., 152, 155–57; denying an enemy the right

to use some seas and sea control, 13; naval bombardment and projection of power ashore, 15; riverine operations during, 118; Tet Offensive, 156

vision statement/message: broad and enduring themes, 49–50; characteristics of, 46–52; defense themes, 49; how to get across, 45–46; inner jointness and, 48–49, 50–51; leadership role in, 47, 50, 51–52; naming the message, 46–47; ownership and protection of, 50–51; packaging and marketing, 47, 50; simplicity of, 46–52; single vision, importance of, 46, 48–49, 51, 52; timing of, 51–52

war: checklist on things to do to win, 1–2; control over enemy as aim of, 120, 121, 124–26; land battles as decisive actions, 37–38; moral forces, strategic value of, 91, 98–100; naval engagements, decisiveness of, 37–38; prevention of, 2, 8, 115, 117; smoldering crisis and outbreak of, 114; winning without fighting, 8, 158

War of 1812, 29

war plan, 169–72; lines of communication, 171; lines of operation, 170–71; maritime communications, 171–72; objective, 170; system of operations, 169–70

War Plan Orange, 47, 48, 60

weapons/artillery: arms and revolution, 130, 134; close air support operations, 17; counter-air/anti-air operations, 16, 17–18; explosives in war, 130, 132–34. *See also* missiles

World War I: amphibious assault operations during, 15; Austrian offensive against Italy, 92, 93–94; blunders and unsound strategy for, 88–96; British army contribution to, 77, 78; British entrance into and character of war, 77, 82–84; British-French coalition during, 78–79, 94; Dardanelles campaign, 77, 78–79, 92; end of, 85–86; German fleet for, 83, 89–90, 92; Jutland battle, 11–12, 94; lessons from, 99–101; as maritime war, 4, 77–86; Marne battle, 91; naval power following, 35; prewar and wartime planning, 78, 89–91; Russian military collapse, 95–96; Schlieffen plan of war, 89, 90–91; sea control and naval strategy during, 11–12, 78–79; sea power and siege of Central Powers, 91–92; Somme battle, 85, 94; start of, 79–81; submarine warfare during, 84–85, 92, 94–95, 96, 97; trench warfare and Western Front operations, 4, 76, 77, 85; U.S. entrance into, 84–85, 88–89, 95–101; Verdun, 93; writings about, 88

World War II: amphibious assault operations during, 15, 23; bombing of Japan, 151–52, 154–55, 160n12; British strategy for, 83–84; coalition for winning, 84; Japanese surrender and end of, 152, 155; Japanese war plan, 154–55; naval power following, 35; naval tactical air operations during, 15; Pearl Harbor attack, 114, 155; sea control and carrier-based air power, 12; sea control and the Pacific War, 123, 125; status of Navy after, 3; submarines and planning for, 12; success of Navy during Pacific War, 3; U.S. victory in Pacific, 152, 154–55

Wylie, J. C., Jr., 120–40

Zeus, 189

Zumwalt, Elmo, 46

ABOUT THE EDITOR

Thomas J. Cutler has been serving the U.S. Navy in various capacities for more than fifty years. The author of many articles and books, including several editions of *The Bluejacket's Manual* and *A Sailor's History of the U.S. Navy*, he is currently the director of professional publishing at the U.S. Naval Institute and Fleet Professor of Strategy and Policy with the Naval War College. He was awarded the William P. Clements Award for Excellence in Education (military teacher of the year) at the U.S. Naval Academy and is a winner of the Alfred Thayer Mahan Award for Naval Literature, the U.S. Maritime Literature Award, and the Naval Institute Press Author of the Year Award.

The Naval Institute Press is the book-publishing arm of the U.S. Naval Institute, a private, nonprofit, membership society for sea service professionals and others who share an interest in naval and maritime affairs. Established in 1873 at the U.S. Naval Academy in Annapolis, Maryland, where its offices remain today, the Naval Institute has members worldwide.

Members of the Naval Institute support the education programs of the society and receive the influential monthly magazine *Proceedings* or the colorful bimonthly magazine *Naval History* and discounts on fine nautical prints and on ship and aircraft photos. They also have access to the transcripts of the Institute's Oral History Program and get discounted admission to any of the Institute-sponsored seminars offered around the country.

The Naval Institute's book-publishing program, begun in 1898 with basic guides to naval practices, has broadened its scope to include books of more general interest. Now the Naval Institute Press publishes about seventy titles each year, ranging from how-to books on boating and navigation to battle histories, biographies, ship and aircraft guides, and novels. Institute members receive significant discounts on the Press's more than eight hundred books in print.

Full-time students are eligible for special half-price membership rates. Life memberships are also available.

For a free catalog describing Naval Institute Press books currently available, and for further information about joining the U.S. Naval Institute, please write to:

Member Services
U.S. NAVAL INSTITUTE
291 Wood Road
Annapolis, MD 21402-5034
Telephone: (800) 233-8764
Fax: (410) 571-1703
Web address: www.usni.org